Recent Results in Cancer Research 59

Fortschritte der Krebsforschung
Progrès dans les recherches sur le cancer

Selective Heat Sensitivity of Cancer Cells

Edited by

A. Rossi-Fanelli R. Cavaliere
B. Mondovì G. Moricca

With 99 Figures

Springer-Verlag
Berlin Heidelberg New York 1977

Prof. ALESSANDRO ROSSI-FANELLI
Institute of Biological Chemistry, University of Rome,
and Center of Molecular Biology C.N.R.,
Rome (Italy)

Prof. RENATO CAVALIERE
Regina Elena Institute for Cancer Research,
Rome (Italy)

Prof. BRUNO MONDOVÌ
Institute of Applied Biochemistry, University of Rome,
Rome (Italy)

Prof. GUIDO MORICCA
Regina Elena Institute for Cancer Research,
Rome (Italy)

Sponsored by the Swiss League against Cancer

ISBN-13:978-3-642-81082-4 e-ISBN-13:978-3-642-81080-0
DOI: 10.1007/978-3-642-81080-0

Library of Congress Cataloging in Publication Data. Main entry under title: Selective heat sensitivity of cancer cells. (Recent results in cancer research; 59). Bibliography: p. Includes index.
1. Cancer. 2. Cancer cells. 3. Thermotherapy. 4. Heat-Physiological effect. I. Rossi-Fanelli, A. II. Series. [DNLM: 1. Heat-Therapeutic Use. 2. Neoplasms-Therapy. 3. Isolation perfusion. 4. Antibiotics, Antineoplastic-Therapeutic use. 5. Radiotherapy. W1 RE106P v. 59/QZ266 S463] RC261.R35 vol. 59 [RC271.T5] 616.9'94'008s [616.9'94'07] 76-48045.

Softcover reprint of the hardcover 1st edition 1977

2125/3140-543210

Preface

Since the first observations of Busch in 1866, the possible use of heat as a therapeutic agent in the cure of cancer has been repeatedly subject to bursts of interest, almost invariably followed by periods of neglect and skepticism. In 1963—1964, this problem was again attacked by us both from the biochemical and from the clinical points of view. The first results of this joint effort were positive beyond expectation, and generated a new revial of studies aimed at the identification of the nature of the biochemical lesion as well as at the optimization of technique and of the therapeutic schedules connected with clinical use.

Although the number of mammalian tumors which have been proved to be heat-sensitive is now relatively large, and although in some cases a correlation has been demonstrated between tumorigenicity and heat-sensitivity of in vitro cultured cell lines, the question of a direct and constant relationship between neoplastic character and higher sensitivity to hyperthermic exposure is still open to continuing investigation and reappraisal.

Several studies deal in fact with the determination of the conditions under which, in vitro and or in vivo, different tumors are efficiently damaged by elevated temperatures. In most cases, the problem involves mainly technical difficulties in reaching the desired temperature in the whole tumor; sometimes instead it appears that the threshold between normal and neoplastic cells is, even in vitro, not as clear-cut as would be desirable for clinical application. The in vitro demonstration, by biochemical techniques, of the heat-sensitivity of each single tumor appears therefore to be a prerequisite for any clinical trial.

On the other hand, in vitro studies have been of great importance in suggesting modifications and extensions of the general scheme of hyperthermic treatment. It has,

e. g., been demonstrated that there is a primary involvement of the processes of nucleic acid biosynthesis. This has led, most recently, to investigation of the possibility of increasing, through association with hyperthermia and therefore by inhibiting repair processes, the therapeutic efficiency of such treatment as ionizing radiation and alkylating drugs, which cause specific damage to nucleic acids. The in vitro results have thus far been most encouraging. From the clinical point of view, however, if the greater efficacy of X-rays after hyperthermic exposure seems to be well ascertained, some divergencies exist among the various authors concerning the long-term validity of hyperthermia plus antiblastic drug combination, the beneficial effects of the latter being sometimes overcome by the loss of the host's ability to build up his own immune defenses against tumor recurrence and/or metastases.

The aim of the present volume is not so much to present the advantages of the use of hyperthermia in the treatment of cancer, but rather to indicate, in a field which is still rapidly evolving, the evidence that has so far built up in favor — or even against — this approach. Due to the fluidity of the current status of research in this field, it has seemed reasonable to ask different authors working in this field to present the experimental facts in the light of their own experience, leaving to them the responsibility for their own views. This approach may lead to omissions as well as to repetition, and even to conflicting opinions on problems still open to debate. We hope nevertheless that, rather than generating confusion, the plurality of points of view may give the reader a more lively idea of some of the problems that remain to be solved. In particular, the chapters dealing with clinical applications reflect essentially the personal experience of the authors, so that the dilemma between regional and total-body hyperthermia is probably overemphasized. Thes reflects, however, a real duality of approach, due to the fact that, in the clinical management of human patients, technical problems, often solved in a rather empirical way, assume a most critical relevance.

ALESSANDRO ROSSI-FANELLI

Contents

List of Contributors

A. F. Anderson, Dept. of Surgery and Cancer Research Laboratory, St. Joseph Hospital Houston, Texas (USA).

A. Bigotti, Regina Elena Institute for Cancer Research, Rome (Italy).

A. Caputo, Regina Elena Institute for Cancer Research, Rome (Italy).

R. Cavaliere, Regina Elena Institute for Cancer Research, Rome (Italy).

F. Colistro, Regina Elena Institute for Cancer Research, Rome (Italy).

C. Crifò, Institute of Biological Chemistry, University of Cagliari (Italy).

J. A. Dickson, Cancer Research Unit, University Department of Clinical Biochemistry, Royal Victoria Infirmary, Newcastle Upon Tyne (England).

B. C. Giovanella, The Stehlin Foundation, Cancer Research Laboratory, St. Joseph Hospital, Houston, Tx 77002 (USA).

A. E. Gutierrez, Department of Surgery and Cancer Research Laboratory, St. Joseph Hospital, Houston, Tx 77002 (USA).

P. D. De Ipolyi, Department of Surgery and Cancer Research Laboratory, St. Joseph Hospital, Houston, Tx 77002 (USA).

C. M. Ludgate, Department of Clinical Surgery, Western General Hospital and University of Edinburgh, Edinburgh (Scottland).

B. Mondovì, Institute of Applied Biochemistry, University of Rome, Rome (Italy).

G. Moricca, Regina Elena Institute for Cancer Research, Rome (Italy).

L. R. Muenz, Clinical and Diagnostic Trials Section, Biometry Branch, National Cancer Institute, Bethesda, Maryland (USA).

R. T. Pettigrew, Western General Hospital, Edinburgh Clinical Teaching Staff, University of Edinburgh, Edinburgh (Scottland).

A. Rossi-Fanelli, Institute of Biological Chemistry, University of Rome, and Center for Molecular Biology, C.N.R., Rome (Italy).

J. S. Stehlin, Jr., Department of Surgery and Cancer Research Laboratory, St. Joseph Hospital, Houston, Tx 77002 (USA).

R. Strom, Institute of Biological Chemistry, School of Medicine, L'Aquila (Italy).

Chapter 1
Introduction

B. C. GIOVANELLA and B. MONDOVI

Before we begin this introduction, let us attempt to dispel all ambiguities about the terms used throughout this book.

Hyperthermia or supranormal temperatures refers, unless specifically stated otherwise, to temperatures which are above the normal body temperature of the animal being studied. In a cell culture system, our reference point is the normal body temperature of the animal from which the cells were derived. This term also implies an upper limit, which is below the temperature where acute damage is produced to cells and organs. For most mammals, this range is between 37 and 45°C. The majority of the studies mentioned here utilize a much narrower range, between 40 and 43°C. Temperatures above 43°C damage normal structures so fast that it becomes impractical to use them to selectively kill tumor cells.

A selective lethal effect of supranormal temperatures on cancer cells in vivo and in vitro has been reported by many investigators during the past hundred years. (For an extensive review of the literature, see (1, 2).)

Most of these studies, performed in vivo, suggested such an effect by demonstrating disappearance or regression of malignant tumors with little damage to the neighboring tissues when the whole body of the host (human or experimental animal) (3-5) or the region containing the tumor (1, 6, 7) are kept at a supranormal temperature for a sufficient length of time. Recently, in vitro studies performed under rigorous experimental conditions have demonstrated that neoplastic cells are more sensitive to heat than normal cells of the same histologic type, even if the latter divide at a faster rate than the tumor cells (8).

This higher thermosensitivity of tumor cells has been demonstrated with both mouse (8) and human cells (9). Accumulating evidence suggests it to be a general property of the neoplastic cell, acquired together with the malignant transformation (10).

Unfortunately, the size of the difference in thermosensitivity between normal and neoplastic cells is not great (8, 11). This small differential makes difficult therapeutic measures based on the thermosensitivity of cancer cells, and has limited their practical application mostly to tumors of the limbs. Fortunately, it seems possible to increase such differential by combining heat and chemotherapeutic agents or heat and ionizing radiations (12-15). This approach has already increased the life expectany of human patients with locally disseminated malignant melanoma of the limbs (9).

Obviously the final goal for therapy is the treatment of the whole body of the patient with disseminated disease. Different methods can be used to raise the temperature of the human body for any length of time. The methods employed are at present insufficiently studied and tested. This stems from the fact that whole-body hyperthermia after a period of rather rapid development in the 1930s when it was used to treat syphilis, gonorrhea, and other infectious diseases, disappeared abruptly from the therapeutic scene after the advent of the antibiotics.

Physiologic studies on the origin, significance, and effects of high temperatures on the animal body waned when its practical application became obsolete. High fevers induced by infectious processes were also rendered rare events by antibiotic treatment (typhoid fever, septic infections, etc.) and prophylactic measures (malaria, erysipelas). When the fever practically disappeared, interest in its mechanism, significance, and effect on the host organism vanished.

After the 1930s, the number of studies on these subjects diminished to a trickle. This lack of experimentation has left us with large gaps in our knowledge and understanding of even such elementary questions as what causes the death of an animal or man subjected to prolonged hyperthermia. We are also ignorant of the effects of fever on the immunologic status of the patient.

Nevertheless, an answer to the first question may very well be the key to the safe application of generalized hyperthermia to cancer patients. The answer to the second may explain the strange finding that destruction by hyperthermic treatment of a tumor in the limb increases the life expectacy of the patient more than amputation of the affected extremity (9).

The most common misconception in this field is that hyperthermia, if too intense (above 42.5°C) or too prolonged, is fatal because of damage to the central nervous system. The fact is that nervous tissue is very thermoresistant. (It can tolerate in vitro prolonged exposure to 48°C without loss of viability (16).) The basis for this fallacy is the constant finding of pathologic lesions in the brains of people who died of heat stroke or hyperpyrexia. Such lesions are identical to those observed in humans who died of asphyxia (17, 18). The reason for this is, simply, that hyperpyrexia victims also die of asphyxia, caused by the paralysis of the respiratory muscles due to accumulation of lactic acid, and saturation of the buffering ability of the plasma (19, 20).

As for the influence of fever on the immune status of the host, it is well known that after infectious diseases, there is a high titer of antibodies against the specific agent which caused the disease and very low titers against other antigens. In effect, the convalescent patient is hyperimmune with respect to the agent of his or her disease (convalescent serum has been used therapeutically with good results), and hypoimmune with respect to any other disease. It has been observed in rabbits that high fever brings first a lymphocytopenia, followed, as the fever abates, by lymphocytosis (21).

Putting together these two facts, it is possible to theorize that the febrile status has evolutionarily developed as a means of "focusing" the immune response of an individual on a specific antigen, simultaneously increasing its size by many orders of magnitude.

In other words, at any given moment, a normal individual has many populations of lymphocytes, each producing a small amount of a different antibody. The acute infection requires the rapid production of one type of antibody. The high fever may be considered as a means of eliminating the majority of the old lympocytes, leaving room for a single new population which matures in the presence of large amounts of an antigen (the agent of the disease), becoming sensitized to it an producing specific antibodies against it. In effect, diseases which produce little or no fever are, in general, of longer duration than febrile ones.

Much more study is necessary in this field before such questions can be answered which certainty and it is regrettable that such an important and widespread phenomenon as fever has attracted so little attention in recent years.

By what mechanism does heat preferentially affect tumor cells? We do not, at present, have a good answer to this question either.

However, two simple but very important facts are firmly established: (I) hyperthermia up to 43°-44°C has a greater effect on dividing cells than on cells that do not divide (13, 22, 23).

If the same cell is exposed to heat in the resting phase and when going through the cell cycle, the heat will be much more harmful to the cycling cell (24, 25). Additional studies in this area have demonstrated that within the cell cycle an area of increased thermosensitivity exists, which covers the late S phase and possibly part of G2 (26-28).

(II) Cell respiration is one of the processes to be affected by hyperthermia (1, 29, 30). This finding clearly differentiates the action of heat on the cell from the action of x-rays, which, even in very high doses, do not effect respiration but do inhibit glycolysis (31), whereas heat generally does not affect glycolysis but does interfere with respiration. Also, hypoxic cells survive x-rays more than oxygenated cells, but are more sensitive to hyperthermia (32). This leads investigators to suspect that hyperthermia and ionizing radiations would act synergistically. Experimentally, this has proved to be true (33), as already been found with chemotherapeutic agents (13).

As for specific biochemical lesions, none so far has been found to be the determining factor in the cell death caused by heat.

Cell membranes are certainly affected by hyperthermia as it is demonstrated by alterations of their permeability and the increased antigenicity of heated cells (34, 35). Biosynthetic processes are also deeply affected by heat. DNA, RNA, and protein synthesis are inhibited, even after relatively short exposures to supranormal temperatures (36-38). This inhibition is reversible but becomes irreversible if the exposure is prolonged further. The effect of heat on the biosynthesis of nucleic acids is of special interest because it seems to be specific for tumor cells (36). The main difficulty in such studies is to establish if the biochemical alterations observed are the cause or the effect of cell death. The finding of temperature-sensitive cell mutants having specific enzymes that are more thermosensitive than those of the parental strain (39, 40), presents us with systems where the problem has already been solved. It remains to be seen if such cases are representative of a general phenomenon. It would be convenient to attribute the thermal death of a cell to the damage of an essential enzyme, and the increased vulnerability of tumor cells to heat to the appearance of a more thermosensitive enzyme. From a practical standpoint, if it is possible to demonstrate a causal link between some specific biochemical damage and irreversible cell division block or cell lysis, then it would be feasible to potentiate the action of heat by the addition of appropriate chemicals.

Another aspect of hyperthermia which is still insufficiently studied is dosage and timing. So far we have very little information on such basic question as what is the time-temperature relationship in cell killing. Only in a few in vitro systems has it been established how many hours at, say 41°C, are equivalent to 1 hr. at 42°C in cell-killing effect (23, 26, 41). We do not know the optimum temperature and time of exposure for the selective killing of tumor cells or how to fractionate the heat treatment for maximum effectiveness.

By now we are fairly certain that sublethal exposures to heat synchronize, to some extent, tumor cells, even in vivo (42, 43). We also know that certain phases of the cell cycle are more thermosensitive then others (26, 27, 28). It is evident that two heat exposures of, say 1 h at 42°C could have a very different killing effect depending on when the second is applied - during the thermosensitive phase or during a refractory period. Probably, this is the cause of the finding that mouse leukemic cells after exposure to 42°C for a short period of time, which is totally insufficient to kill them, are killed by a further short exposure to heat even at a temperature as low as 40°C, which alone could not appreciably affect such cells if applied continuously for many hours (13).

The future of hyperthermic treatment of human tumors is uncertain. What is certain is that at least some human tumors can be successfully treated by hyperthermia alone or in combination with chemotherapeutic drugs, and that this field requires and deserves more studies and further efforts.

REFERENCES

1. CAVALIERE, R., CIOCATTO, E.C., GIOVANELLA, B.C., HEIDELBERGER, C., JOHNSON, R.O., MARGOTTINI, M., MONDOVI, B., MORICCA, G., ROSSI-FANELLI, A.: Selective heat sensitivity of cancer cells. Cancer (Philad.) 20, 1351 (1967)
2. SUIT, H.D., SHWAYDER, M.: Hyperthermia: Potential as an anti-tumor agent. Cancer (Philad.) 34, 122 (1974)
3. WARREN, S.L.: Preliminary study of the effect of artificial fever upon hopeless tumor cases. Amer. J. Roentgenol. 33, 75 (1935)
4. DICKSON, JA. A., MUCKLE, D.S.: Total body hyperthermia versus primary tumor hyperthermia in the treatment of the rabbit VX-2 carcinoma. Cancer Res. 32, 1916 (1972)
5. PETTIGREW, R.T., GALT, J.M., LUDGATE, C.M., SMITH, A.N.: Clinical effects of whole-body hyperthermia in advanced malignancy. Brit. med. J. 4, 679 (1974)
6. WESTERMARK, N.: The effect of heat upon rat tumors. Scand. Arch. Physiol. 52, 257 (1927)
7. HALL, R.R., SCHADE, R.O.K., SWINNEY: Effects of hyperthermia on bladder cancer. Brit. med. J. 2, 593 (1974)
8. GIOVANELLA, B.C., MORGAN, A.C., STEHLIN, J.S., WILLIAMS, L.J.: Selective lethal effect of supranormal temperatures on mouse sarcoma cells. Cancer Res. 33, 2568 (1973)
9. STEHLIN, J.S., GIOVANELLA, B.C., de IPOLYI, P.D., MUNTZ, L.R., ANDERSON, R.F.: Results of hyperthermic perfusion for melanoma of the extremities. Surg. Gynec. Obstet. 40, 339 (1975)
10. CHEN, T.T., HEIDELBERGER, C.: Quantitative studies on the malignant transformation of mouse prostate cells by carcinogenic hydrocarbons in vitro. Int. J. Cancer 4, 166 (1969)
11. BENDER, E., SCHRAMM, T.: Untersuchungen zur Thermosensibilität von Tumor- und Normalzellen in vitro. Acta biol. med. germ. 17, 527 (1966)
12. STEHLIN, J.S.: Hyperthermic perfusion with chemotherapy for cancer of the extremities. Surg. Gynec. Obstet. 129, 305 (1969)

13. GIOVANELLA, B.C., LOHMAN, W.A., HEIDELBERGER, C.: Effects of elevated temperatures and drugs on the viability of L1210 leukemia cells. Cancer Res. 30, 1623 (1970)
14. MUCKLE, D.S., DICKSON, J.A.: Hyperthermia (42°C) as an adjuvant to radiotherapy and chemotherapy in the treatment of the allogenic VX-2 carcinoma in the rabbit. Brit. J. Cancer 27, 307 (1973)
15. YERUSHALMI, A., HARKEDAR, I.: Enhancement of radiation effects by heating of the tumor. Israel J. med. Sci. 10, 772 (1974)
16. MOSSA, S.: Ulteriori studi sulla velocita di accrescimento dei neuriti coltivati in vitro in funzione della temperatura ambiente. Arch. exp. Zellforsch. 4, 188 (1927)
17. HARTMAN, F.W.: Lesions of the brain following fever therapy, etiology and pathogenesis. J.A.M.A. 109, 2116 (1937)
18. GORE, I.: The pathology of hyperpyrexia. Amer. J. Path. 25, 1029 (1949)
19. HALL, W.W., WAKEFIELD, E.G.: A study of experimental heat-stroke. J.A.M.A. 89, 177 (1927)
20. FRANKEL, H.M., ELLIS, J.P.Jr., CAINS, S.M.: Development of tissue hypoxia during progressive hyperthermia in dogs. Amer. J. Physiol. 205, 733 (1963)
21. DOAN, C.: Peripheral blood phenomena and differential response of bone marrow and lymph nodes to hyperpyrexia. Radiology 30, 382 (1938)
22. GIOVANELLA, B.C., HEIDELBERGER, C.: Biochemical and biological effect of heat on normal and neoplastic cells. Proc. A.A.C.R. 9, 24 (1968)
23. PALZER, R.J., HEIDELBERGER, C.: Studies on the quantitative biology of hyperthermic killing of HeLa cells. Cancer Res. 33, 415 (1973)
24. GIOVANELLA, B.C., MOSTI, R., HEIDELBERGER, C.: Further studies on the lethal effect of heat on tumor cells. Proc. A.A.C.R. 10, 29 (1969)
25. DICKSON, J.A., SHAH, D.M.: The effects of hyperthermia (42°C) on the biochemistry and growth of a malignant cell line. Europ. J. Cancer 8, 561 (1972)
26. WESTRA, A., DEWEY, W.C.: Variation in sensitivity to heat shock during the cell cycle of Chinese hamster cells in vitro. Int. J. Radiat. Biol. 19, 467 (1971)
27. DEWEY, W.C., WESTRA, W., MILLER, H.H., NAGASAWA, H.: Heat-induced lethality and chromosomal damage in synchronized Chinese hamster cells treated with 5-Bromodeoxyuridine. Int. J. Radiat. Biol. 20, 505 (1971)
28. PALZER, R.J., HEIDELBERGER, C.: Influence of drugs and synchrony on the hyperthermic killing of HeLa cells. Cancer Res. 33, 422 (1973)
29. MONDOVI, B., STROM, R., ROTILIO, G., FINAZZI AGRO, A., CAVALIERE, R., ROSSI-FANELLI, A.: The biochemical mechanism of selective heat sensitivity of cancer cells. I. Studies on cellular respiration. Europ. J. Cancer 5, 129 (1969)
30. MUCKLE, D.S., DICKSON, J.A.: The selective inhibitory effect of hyperthermia on the metabolism and growth of malignant cells. Brit. J. Cancer 25, 771 (1971)
31. CAPUTO, A., GIOVANELLA, B.C.: The action of ionizing radiations on the respiration and on the aerobic and anaerobic glycolysis of Ehrlich mouse ascites cells. Radiat. Res. 13, 809 (1960)
32. HAHN, G.M.: Metabolic aspects of the role of hyperthermia in mammalian cell inactivation and their possible relevance to cancer treatment. Cancer Res. 34, 3117 (1974)
33. BEN-HUR, E., BRONK, B.V., ELKIND, M.M.: Thermally enhanced radiosensitivity of cultured Chinese hamster cells. Nature (New Biol.) 238, 209 (1972)

34. STROM, R., CAIAFA, P., MONDOVI, B., ROSSI-FANELLI, A.: Temperature dependence of transcellular migration of glutamate in normal and neoplastic cells in: Permeability and function of biological membranes. Eds.: Keynes, R.D., Loewenstein, W.R., Pethica, B.A., North-Holland Publishing Co., 1970
35. MONDOVI, B., SCIOSCIA SANTORO, A., STROM, R., ROSSI-FANELLI, A.: Increased immunogenicity of Ehrlich ascites cells after heat treatment. Cancer (Philad.) 30, 885 (1972)
36. MONDOVI, B., FINAZZI AGRO, A., ROTILIO, G., STROM, R., MORICCA, G., ROSSI-FANELLI, A.: The biochemical mechanism of selective heat sensitivity of cancer cells. II. Studies on nucleic acids and protein synthesis. Europ. J. Cancer 5, 137 (1969)
37. McCORMICK, W., PENMAN, S.: Regulation of protein synthesis in HeLa cells: Translation at elevated temperatures. J. molec. Biol. 39, 315 (1969)
38. WAROCQUIER, R., SCHERRER, K.: RNA metabolism in mammalian cells at elevated temperature. Europ. J. Biochem. 10, 362 (1969)
39. THOMPSON, L.H., HARKINS, J.L., STRANNERS, C.P.: A mammalian cell mutant with a temperature-sensitive leucyl-transfer RNA synthetase. P.N.A.S. 70, 3094 (1973)
40. SHARP, J.D., CAPECCHI, N.E., CAPECCHI, M.R.: Altered enzymes in drug-resistant variants of mammalian tissue culture cells. P.N.A.S. 10, 3145 (1973)
41. HARRIS, M.: Growth and survival of mammalian cells under continuous thermal stress. Exp. Cell Res. 56, 382 (1969)
42. MARTIN, R.J., SCHLOERB, P.R.: Induction of mitotic synchrony by intermittent hyperthermia in the Walker 256 rat carcinoma. Cancer Res. 24, 1997 (1964)
43. SAPOZINK, M.D., DESCHNER, E.E., HAHN, E.W.: Induction of mitotic synchrony by intermittent hyperthermia in Ehrlich carcinoma in vivo. Nature (Lond.) 244, 299 (1973)

Chapter 2
Biochemical Aspects of Heat Sensitivity of Tumour Cells

R. STROM, C. CRIFO, A. ROSSI-FANELLI, and B. MONDOVI

1. INTRODUCTION

As mentioned in the Introduction, and as shall be taken up in the following chapter, several lines of evidence have indicated, starting from the first observation by BUSCH (17) over a century ago, that temperatures above the physiologic value might have destructive effects on cancer. As shall be described in the next chapter, *in vivo* therapeutic applications require the solution of a number of problems: validity and verification, in every case, of the assumption that hyperthermia will be beneficial, i.e., that tumor cells will be more temperature-sensitive than the surrounding normal ones; decisions as to the level of hyperthermia that must be reached, as well as its duration, and the rate at which the final level is to be attained; possible use of substances that act as potentiators of hyperthermia, or of procedures which, associated with the high temperature, result in a more selective and/or extensive damage of the tumor, or that limit the danger of metastatic diffusion; evaluation of the immune status of the patient, in view of the possibility (which shall be discussed in this chapter and elsewhere in this volume that immunity plays a role in tumor regression and prevention of metastasis; technical difficulties in maintaining, with minimal oscillations and for the time required, a given high temperature in the whole tumor mass, controlling it constantly and avoiding the appearance in the patient of adverse effects, such as heat stroke, fluid and/or electrolyte loss, shock, acidosis, renal failure, local edema, etc.; subsequent treatment of the patient in order to avoid not only recurrences and/or metastases but also, as has sometimes been the case in large tumor masses undergoing rapid necrosis (19), late complications due to resorption of heat-damaged tumor tissues.

Some of these problems are mainly technical, in that they are strictly related to the type, the dimensions, and the position of the tumor and to the kind of hyperthermic treatment adopted. Other aspects have wider biological implications, and require, for a better understanding, the use of *in vitro* investigations. In this chapter we shall discuss the results of these investigations, along the following main lines:

1. Are we justified in speaking of a generalized higher heat sensitivity of tumor cells?

2. Which aspects of the structure and/or metabolism of tumor cells are damaged by exposure of tumor cells to supranormal temperatures?

3. Which target is likely to be the critical one, or, at least the most sensitive one?

Two aspects of heat treatment, which are of utmost importance for in vivo applications, shall be mentioned and discussed, but only briefly. The concern the danger, upon interruption of thermal exposure, of a recovery of the partially damaged tumor cells (which the possible appareance and/or selection of temperature-resistant clones), and the comparison of heat treatment with other cytocidal treatment, especially with respect to a possible synergic action. Of particular interest from this point of view is the possible involvement of the immune system in the therapeutic effectiveness of heat treatment.

2. SELECTIVITY OF THE LETHAL EFFECT OF SUPRANORMAL TEMPERATURES ON TUMOR CELLS

In several studies (5, 7, 9, 19, 25, 50, 55, 56, 58, 59, 62, 76-78, 89, 92, 108, 111, 112, 120-123, 135, 142, 143, 145, 159, 168, 170, 171, 174), some of which were performed more than 50 years ago, it has been asserted that neoplastic cells are, in vitro as well as in vivo, more easily damaged by supranormal temperatures than are normal cells.

Four main problems, however, are raised by such assertions:

1. What criterion is used to evaluate cell damage?

2. How close are the "tumor" cells to the "normal" cells to which they are compared?

3. Is the higher heat sensitivity, if present, really related to the "neoplastic character" of the tumor cells, rather than to the fact that these tumor cells are usually aneuploid and have an increased proliferation rate?

4. How distant are the temperature levels that damage tumor cells from those that also affect normal cells?

The first question is not as trivial as it may seem. The most reliable method for evaluating damage of neoplastic cells is surely the loss, after heat treatment, of their ability to produce tumors upon inoculation into an appropriate recipient (58, 59). This criterion cannot, however, be applied to normal, nontumorigenic cells. For a comparison, therefore, four possible criteria are available:

1. Appearance of given morphologic changes

2. Loss of cell membrane impermeability, as monitored by entry of certain dyes (trypan blue, eosin, nigrosin) or by efflux of radioactive chromate

3. Inability of further growth under culture conditions

4. More or less complete inhibition of fundamental biochemical activities: respiration, glycolysis, biosynthesis of nucleic acids and/or of proteins.

The morphologic criterion is difficult to use, since, as shall be seen below, the attention of the various authors has been focused on different parts of the cell; moreover, the extent of these morphologic changes is difficult to evaluate quantitatively, and is liable to vary with time.

As for the loss of cell membrane impermeability, it is, because of its simplicity, extremely useful when a fast screening method is required; however, as has been pointed out by several authors (58, 66), and as shall also be shown in Figure 5 of this chapter, the absolute validity of this parameter is to be questioned since it is often subject to variations according to experimental in vitro conditions and because it occurs rather late, so that it should be considered almost as an epiphenomenon.

Similar objections hold also for the loss of ability to resume growth when under appropriate culture conditions, since a relatively long time is required before the response can be evaluated. Moreover, at the single cell level, it is an all-or-none response, requiring therefore a multitude of events to take place and which is, moreover, subject to the influence of recovery processes.

In our judgment, therefore, the inhibition of some fundamental biochemical activities appears to be the best parameter for evaluating the extent of cell damage: a whole range of phenomena can be investigated, the possible accuracy of the investigation being almost unlimited. As shall be seen below, biosynthesis of nucleic acids, especially of DNA and of ribosomal RNA, appears to be most rapidly and severly affected by exposure to high temperatures (162, 168).

Of the authors cited at the beginning of this chapter who have investigated the in vitro heat sensitivity of tumor cells, some (25, 58, 62, 76, 77, 92, 112, 121, 135, 145, 162, 174) have not correlated their observations on tumor cells with similar effects on normal cells, being generally more interested in the description of the phenomena or in the mechanism (1) underlying them. In those studies in which a comparison between normal and tumor cells has been performed, the choice of material selected for the purpose is sometimes open to criticism: it does not in fact appear justified to compare spleen cells to Ehrlich ascites cells (171), nongrowing fibroblasts to actively dividing epithelial cells from carcinoma lines (7, 89), or even normal, resting lymphocytes to leukemic lymphoblasts or myeloblasts (123, 142, 143). However, generally, and with the notable exception of cells derived from the Rous sarcoma of fowls (55, 56), it was observed that, under similar culture or suspension, normal cells are more thermoresistant than neoplastic cells (9, 170).

In a few cases the normal and the neoplastic cells studied were strictly related: proliferating connective tissue cells were compared to sarcoma cells (88), hepatocytes from normal or regenerating liver to cells from fast-growing and/or minimal deviation hepatomas, normal mouse prostate cells in culture to the same cells in vitro transformed by carcinogenic hydrocarbons (120), normal mesenchymal embryonic cells in culture to a subline of the same cells which had spontaneously acquired tumor-producing ability, or to cells derived from 3-methylcholanthrene-induced sarcomas (59). In all these cases, exposure to a supranormal temperature resulted in a cellular damage which appeared to be slective for the "neoplastic" cells.

Three exceptions should, however, be mentioned. GERICKE (55, 56) showed that the Rous sarcoma of the fowl is extremely thermoresistant, surviving in temperatures as high as 50°C.

A certain degree of heat resistance was present also in some hamster-adapted human carcinoma lines (56); this finding does not, however, affect the validity of the thesis of a generally greater thermosensitivity of neoplastic cells in mammals including man.

LOVE et al. (93), however, failed to find any significant difference in the responses of normal diploid cells (human cell lines, or BHK 21) and of neoplastic cells (HeLa), when subjected to 45-46°C. Such a temperature may, however, be so high as to bring about extensive cell damage in both kinds of cells. GIOVANELLA et al. (59), for example, have shown that in their mouse mesenchymal cells grown in tissue culture, the differential heat sensitivity of tumor cells was highest at 42.5°C, whereas at 43-44°C it already tended to disappear.

A most intriguing report was that by OSSOVSKI and SACHS (117), that 3T3 cells, either "normal" or transformed by 3-methylcholanthrene or by X-irradiation, underwent a more severe inhibition of growth at 41°C than did the same cells transformed by either simian virus 40 or by polyoma virus. It appeared that the effects, if any, of viral transformation in vitro -- which, as a first approach, could be equated to viral carcenogenesis -- were opposite to those reported for the spontaneous or chemically induced transformation of normal cells into neoplastic ones. In similar experiments, however, KACHANI and SABIN (82) found that _in vitro_ transformation of hamster cells by simian virus 40 or by polyoma virus did not modify the thermoresistance of these cells; if instead these virus-transformed cells were injected into hamsters and the cells derived from the tumor thus produced were examined for heat sensitivity, it was observed that they were more thermosensitive than normal hamster embryo cells.

A possible explanation for this discrepancy is that _in vivo_ passage causes a selection of really tumorigenic cells out of a majority of virus-transformed but nontumor-producing ones, and that only tumorigenic cells exhibit a pronounced sensitivity to the high temperature.

As for cells primarily derived from human tumors, for which the question of heat sensitivity is of utmost importance for judging the possible success of hyperthermic treatment, the problem of their thermosensitivity is complicated by the lack of controls, i.e., normal cells which may be assumed to be the normal counterpart of tumor cells. The use of normal resting lymphocytes or leukocates as compared to leukemic lympoblasts or myeloblasts (123, 142, 143) has already been mentioned and criticized. It may, however, be further mentioned that, as a rule, leukemic cells were found to be more heat-sensitive. In our own experience, all cells derived from human tumors (five melanomans, three osteosarcomas, two rectal carcinomas) showed a relevant (> 40%) degree of inhibition at 42-42.4°C, at least when DNA and/or RNA synthesis were followed; when no heat sensitivity could be evidenced (as occurred with two melanomas, one rectal carcinoma, and one soft tissue sarcoma), cytologic examination constantly showed that over 80% of the visible cells extracted from the tumor and present in the suspension had clearly no neoplastic character, having generally the appearance of normal lymphocytes (19, 108), and unpublished observations). As reported in the following chapter, in J.A. DICKSONS's experience only about one-third of the human tumors examined -- out of a relatively large number -- appear to be heat-sensitive, i.e., damaged by a temperature of 42°C. Whether this reflects the real situation, or is due to the experimental culture conditions used and/or the method employed for evaluating cell damage (respiration and/or glycolysis), is still an open question.

The possibility should be considered that the increased thermosensitivity of tumor cells is related either to their aneuploidy, or to the increased rate of growth. In fact, the low thermal tolerance of the specific germinal epithelium of testes is well known (26). An influence of ploidy on thermosensitivity has also been observed in yeast, where diploid cells appear to be 2.5 times less sensitive to the lethal effect of 52°C incubation than haploid ones (187). In mammalian cells under culture conditions, HARRIS could show neither a difference in chromosome number between "wild-type" pig kidney cells and temperature-resistant variants from the same cells (66), nor any influence of ploidy on the frequency of recovery of heat-resistant mutants (60). McBURNEY and WHITMORE (105), however, were able to isolate temperature-sensitive (ts) mutants from diploid CHO cells but not from t traploid ones: this was taken as confirming the fact that these ts mutations behave as genetically recessive (107, 166). No general correlation can, however, be evidenced, in either normal or tumor cells, between degree of aneuploidy and heat sensitivity (59).

As for the possible effects of increased growth rate, a haploid strain of Saccharomyces cerevisiae exhibited different heat sensitivity according to its state: exponentially growing cells were much more sensitive (52°C) than resting cells (140). Also, in mammlian systems, it has been postulated that the increased thermosensitivity of tumor cells is related to the more rapid growth of these cells compared to normal ones (9). Almost 50 years ago, BUCCIANTE (16) found that chick heart fibroblasts in culture were more sensitive to high temperature when dividing than when in the resting state. More recently, working with "normal" diploid fibroblasts, LEVIN and ROBBINS (89) found that these cells were not affected at all by exposure to 42°C as long as they were in the resting state; in the growing state, however, they strongly resented from this high temperature, being almost as damaged as the heteroploid "cancer" cells. An analogous phenomenon was found by DICKSON and SHAH (35) to occur also in monolayer cultures of a malignant cell line: cells in a logarithmic phase of growth were more susceptible to hyperthermic damage than cells with a minimal replication rate. Such a dependence would be justified by the fact, which shall be discussed below, that thermosensitivity is not equally distributed throughout the cell cycle, the S and possibly the M phase being several times more sensitive than the G1 and G2 phases. GIOVANELLA et al. (59), however, found no correlation between growth rate and heat sensitivity of a number of normal mouse embryo fibroblasts or mouse sarcoma cells.

In conclusion, there is strong evidence indicating that mammalian neoplastic cells -- with the possible exception of those derived from a viral transformation -- are more thermosensitive than normal ones. Whether this could be considered an intrinsic property of the "neoplastic character" is questionable, because neoplastic transformation often involves changes of cell metabolism and physiology, which might by themselves cause the increase in heat sensitivity. Experiments in which similar cells with comparable growth rates have been considered, e.g., minimal deviation hepatoma vs. hepatocytes from regeneration liver (108, 111), or "normal" cell lines before and after in vitro transformation to tumorigenic cells (59, 120), indicate it may be justified to speak of an increased thermal sensitivity of tumor cells.

Two relevant points here concern the temperature at which damage occurs and its irreversibility. As a general rule, it has been found that the cell damage, as measured by a veriety of techniques, occurs in "tumor" cells at temperatures which are not only lower than those affecting "normal" ones but which are often in the range between 41 and 44°C (5, 19, 55, 56, 59, 76, 78, 88, 89, 108, 111, 112, 120, 121, 143, 145, 159, 171, 174), i.e., in a region of supranormal temperatures which can be attained in vivo without damage to the surrounding normal tissues of the patient or experimental animal. At a given temperature, there is a sigmoid dependence of the surviving fraction upon time of exposure to heat (67, 121, 175). The rate of cell death, as calculated from the slope of the central part of the sigmoid curve, is higher for higher temperatures (58, 59, 175); in some cases its logarithm is, in the range considered, proportional to the reciprocal of the absolute temperature, as in Arrhenius plot (174, 175). The exposure times required for a total or subtotal damage to the cell population will therefore vary with the temperature, as will be seen in Figure 18 of the next chapter. In some cases (93, 106, 170, 172), especially with long-established "tumor" cell lines (e.g., HeLa cells), the damage was found to be reversible, and therefore comparable to that occurring under certain conditions of heat exposure also in "normal" cell lines (2, 93, 147, 149). In most other cases, however, the cell damage was irreversible, and only occasionally could a certain degree of recovery be shown (121, 145).

3. NATURE OF THE LESIONS CAUSED BY SUPRANORMAL TEMPERATURES IN MAMMALIAN CELLS

Upon exposure to supranormal temperatures of mammalian tumor cells -- and sometimes also of "normal" cells in culture -- a number of morphologic and biochemical modifications occur. Although the assumption of a unique mechanism for the hyperthermic killing of different cell types (both normal and neoplastic, exhibiting different degrees of thermosensitivity) is not per se justified, the results obtained by various authors rarely contradict each other, and are therefore susceptible of mutual comparison and common discussion.

From the morphologic point of view, exposure to supranormal temperature brings about, in heat-sensitive cells, the appearance of "bizarre shapes, surface blebbing and a striking increase in nuclear size" (89). After a relatively short heat treatment, mitochondria, Golgi and endoplasmic reticulum vesicles and lysosomes appear more or less indistinguishable from those of untreated cells (70, 89), except, in some cases, for an increase in the number of lysosomes (70). Polyribosomes are either absent (70), or, if present morphologically, cannot be revealed by sucrose gradient centrifugation techniques (89, 106). In the nucleus, chromosomal alterations have been found in some cases (32). However, the main target, from the morphologic point of view, appears to be the nucleolus with retraction of the intranucleolar chromatin and disappearance of the nucleolar reticulum, followed by loss or reduction of the granular nucleolar component (70, 148-150); the nucleolus becomes therefore morphologically homogeneous, being occupied uniquely by its RNA-ase sensitive fibrillar component. In some cases, formation of intensely stained perichromatin granules and accumulation of interchromatin granules can also be shown (70).

From the biochemical point of view, several aspects of cell metabolism have been shown to be affected by thermal exposure. WESTERMARK (174) found that glycolysis, both aerobic and anaerobic, was rapidly inhibited in two experimental rat tumors, the thermal sensitiveness appearing to depend on the glycolytic capability of the tissue.

In our hands (19, 111) glycolysis was, in all tumors examined (Novikoff rat hepatoma, Morris 5123 minimal deviation rat hepatoma, Yoshida ascites rat sarcoma, Ehrlich ascites mouse carcinoma, seven different human melanomas, three different human osteosarcomas, one human rectal carcinoma), unaffected by exposure to temperatures up to 44°C. In a few cases, where the rate of lactate production seemed to decline upon prolonged thermal exposure, it was subsequently ascertained that the inhibition observed could be attributed to an acidification of the medium. Similar results have also been obtained by other authors. VON ARDENNE resports that glycolysis is exhibited in Ehrlich ascites mouse carcinoma cells, but that this inhibition is a very late event (6). In DICKSON's experience (see 35-37, 112, and next chapter) inhibition of glycolysis was often, but not always, a fairly good parameter in evaluating the degree of damage of human was well as of experimental cells;

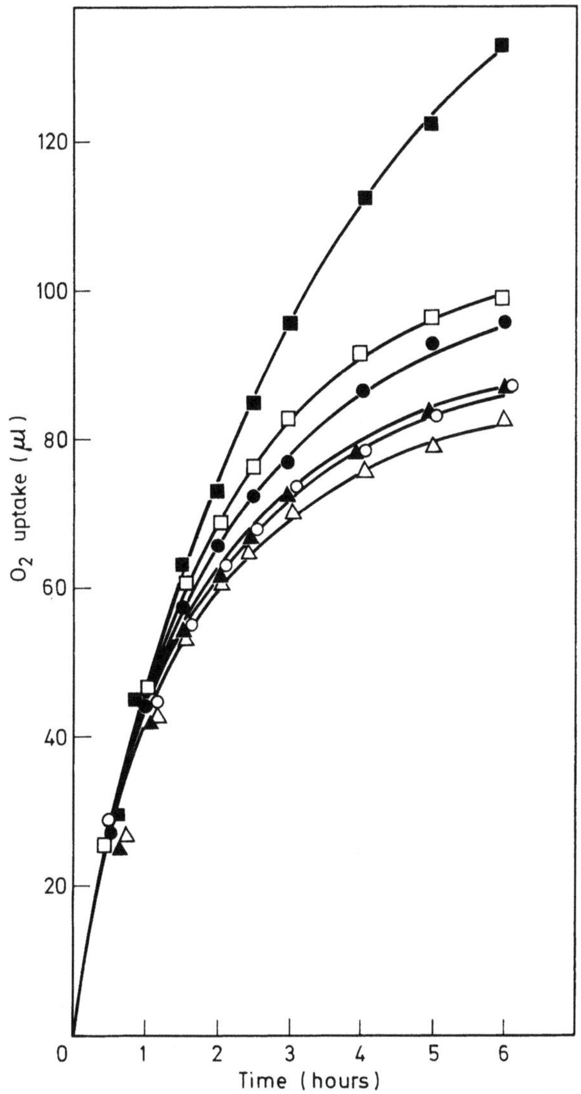

Fig. 1. Oxygen uptake of Novikoff hepatoma cells kept at 43°C for different time intervals and then transferred to 38°C. At time 0, the cells, suspended in Krebs Ringer phosphate solution, were mixed in Warburg flasks with 15 mM glucose and 13 mM succinate, and incubated at 43°C. After given time intervals (■ : immediately; □ : 30 min; ●: 1 h; o: 2 h; ▲ : 3 h; △ : 4 h), during which O_2 uptake was measured, the Warburg flasks were transferred to 38°C, and respiration measured again. Values are referred to 10 mg dry weight of cells

a serious drawback to the validity of this parameter, however, was ascertained by DICKSON and ELLIS (33): anaerobic glycolysis of Yoshida sarcoma cells underwent significant inhibition under conditions of insufficient thermal exposure, which led to enhanced tumor cells dissemination rather than to therapy of the primary tumor.

A much more reliable parameter was oxygen consumption which underwent severe inhibition upon thermal exposure of the neoplastic cells. This was verified for a large number of tumor cells by us (19, 11) as well as by VON ARDENNE's group (6) and by DICKSON and coworkers (33, 35-37, and next chapter). Only in a few cases (e.g., with Ehrlich ascites mouse carcinoma cells below 44°C in our hands, but not according to VON ARDENNE) was the oxygen consumption not affected until extensive cell damage had taken place.

Figure 1 shows, however, that e.g., in Novikoff hepatoma cells, the extent of cell damage following a relatively short (30-60 min) exposure to 43°C is not revealed through a decrease of oxygen consumption until a further interval of time ($\geqslant$ 30-60 min) has elapsed. The inhibition of respiration therefore appears to be a relatively late phenomenon in the sequence of events which follow thermal exposure. Moreover, the extent of inhibition varies with time, and its evaluation may be laborious in tumors with a moderately low thermal sensitivity, such as the minimal deviation Morris 5123 rat hepatoma (19, 111). On the other hand, it reflects essentially the irreversible cell damage which has taken place, and appears therefore to be an extremely useful parameter for a first estimation of the degree of heat sensitivity of a given tumor.

A different set of parameters which are affected by thermal exposure concern the synthesis of macromolecules necessary to the cells: proteins and nucleic acids. In a fest series of results, we found (108) that, in cells from Hovikoff rat hepatoma, as well as in cells from Morris 5123 rat hepatoma and from a human osteosarcoma, these syntheses were all severly inhibited.

Table 1. Comparative effects of preincubation at 42 or 43°C on rates of O_2 consumption nucleic acid synthesis, and protein synthesis at 38°C by Novikoff hepatoma, Morris 5123 minimal deviation hepatoma, and regenerating liver cells

Derivation of cells	Parameter observed	Temperature and duration of exposure			
		42°C, 2 h	43°C, 30 min	43°C 2 h	43°C, 3,5 h
Novikoff	O_2 uptake	15	90-40 (a)	30	15- 20
hepatoma	DNA synthesis	-	28	4	-
RNA synthe	RNA synthesis	-	30-40	15- 30	-
	Protein synthesis	-	-	5- 10	-
Morris 5123	O_2 uptake	100	-	90- 95	70- 75
hepatoma	DNA synthesis	61	-	37	-
	RNA synthesis	-	-	83	-
	Protein synthesis	55	-	40	-
Regenerating	O_2 uptake	100	-	100	100-105
liver	DNA synthesis	-	-	105-115	-
	RNA synthesis	-	-	95	-
	Protein synthesis	-	-	105-120	-

Experimental conditions as in refs. 2, 22, 23. The values are expressed as percent of the activity of analogous preparations kept at 38°C for the same time and under the same conditions. [a]Indicates a progressive increase in the extent of inhibition

Table 1 summarizes the results obtained in these cells, as compared to cells from regenerating rat liver; the cell damage caused by preincubation at an elevated temperature is comparatively evaluated through oxygen consumption or through incorporation of labeled thymidine, uridine, and leucine into DNA, RNA, and proteins, respectively.

It can be seen that biosynthetic processes are, as a rule, more sensitive to thermal exposure than is oxygen uptake. The data shown in Table 1 indicate that uridine incorporation into RNA is somewhat less inhibited than synthesis of either DNA or protein. Further investigation shas shown, however, that this occurs only under certain experimental conditions. In HeLa cells, LEVINE and ROBBINS (89) have shown that DNA synthesis is affected more than RNA synthesis, and the latter more than protein synthesis. In monolayer cultures of a malignant cell line, DICKSON and SHAH (35) have found that RNA synthesis is in fact the parameter most severely affected. Since it appears that the synthesis of ribosomal RNA is preferentially inhibited compared to other kinds of RNA (2, 162, 172), it seems reasonable that differences in the relative rates of synthesis of various RNAs may account for the differences in the experimental results obtained by various authors, and that inhibition of ribosomal RNA synthesis (rather than inhibition of total RNA synthesis) is a very sensitive parameter for evaluating thermal sensitivity.

A deeper insight into the mechanism of hyperthermia-induced cell damage can be obtained by a more thorough investigation of the maturation processes of ribosomal RNA; this aspect shall be treated below.

As for the effects of thermal exposure upon other biosynthetic pathways, neither glucose incorporation into polysaccarides nor choline or linoleic acid incorporation into lipids appear to be affected (89).

4. THERMAL SENSITIVITY AND THE CELL CYCLE

Further information about the mechanism(s) by which tumor cells are damaged by exposure to supranormal temperatures has been obtained through investigations concerning the differential thermal sensitivity of various phases of the cell reproductive cycle. It has already been mentioned that in all kinds of cells examined bacterial or yeast cells (140, 176) as well as "normal" mammalian fibroblasts (16, 89) or "tumor" cells in culture (35) -- there is a general trend toward a higher sensitivity of actively growing cells. On the other hand, the various phases of the cell cycle do not appear to be equally sensitive to thermal exposure. PALZER and HEIDELBERGER (122) showed that HeLa cells in the late S phase were approximately 7 times more sensitive to heat-induced damage than in G1 or G2. This finding, which is similar to the results obtained in yeast (140), was confirmed in Chinese hamster ovary cells by WESTRA and DEWEY (172); these authors indicated, however, the presence of a second peak of high thermal sensitivity in the M phase, thus confirming previous results by SISKEN et al. (152) in human amnion cells. The same authors (32) found that heat treatment in the S phase caused a considerable degree of chromosomal damage, consisting mainly of deletions and chromatic exchanges; the frequency of chromosomal aberrations per cell compared well with the extent of cell killing. The pattern was, however, different in other phases of the cycle: in the G1 and M phases, chromosomal damage was less frequent, and could not account for cell killing (32); on the other hand, heat treatment in the M phase (but not in the other phases!) interfered with the completion of cytokinesis, so that tetraploid cells appeared (32, 175).

As for the effects of heat treatment on the duration of the various phases of the cell cycle, SISKEN et al. (152) have shown that in all cases -- except in anaphase, where there was an all-or-none effect -- exposure to high temperatures caused an elongation of the times necessary for the passage to the next phase; metaphase was, however, by far the most sensitive. It is therefore not surprising that heat treatment, when not lethal per se has been shown to delay the onset of the next generation cycle (35, 121, 175), and that repeated sublethal heat treatments in cell synchronization (104, 137).

In conclusion, the effects of thermal exposure appear to be of two main types, depending on whether the exposure has involved mitosis or not. In the former case, there appear to be essentially a block of cytokinesis, which is probably to be ascribed to denaturation of the spindle proteins (32, 175); the partial heat-resistance of anaphase, which has been reported to be the only phase the duration of which is not affected by heat (152), may probably be ascribed to the all-or-none nature of the spindle damage, so that cytokinesis either is completely stopped or continues at its normal rate. On the other hand, outside of this mitotic phase -- which usually occupies only a short fraction of the cell cycle -- a different kind of cell damage occurs, the highest heat sensitivity being exhibited during the S phase. This second kind of cell damage is probably the prominent one both in living tissues and in nonsynchronized cultures, where the number of cells in actual mitosis is always very low; therefore, in the following paragraphs it shall be the main object of discussion on the nature of heat-induced lesions of tumor cells.

5. POSSIBLE MOLECULAR MECHANISM(S) OF HEAT-INDUCED CELL DAMAGE

A. Nucleic Acids and/or Proteins

The damage caused to tumor cells by thermal exposure occurs within such a limited range of temperatures that it appears plausible that it may be reconduced to a restricted number of lesions, if not to a single one, at the molecular level. Since biosynthetic pathways appear to be most severely affected, the hypothesis of damage to some macromolecular component(s) of the cell should be examined.

a) DNA

Morphologically, the impairement of biosynthetic activities following heat treatment has been found to correspond to structural alterations of the nucleolus and/or the appearance of chromosomal aberrations. Also, several data on cell behavior and physiology upon thermal exposure seem to give support to the hypothesis of nucleic acid involvement. PALZER and HEIDELBERGER (121) found that in HeLa cells there is often, upon thermal exposure, a delayed cell damage, at least one cell division occurring before cell death; this might be taken as indicating a lesion restricted to the genome. On the other hand, in various systems, not only has ploidy been reported to affect, in some cases at least, the degree of thermal sensitivity (see ref. 178, and above), but heat treatment is able to induce mutations in a number of both unicellular and multicellular organisms (114, 140, 182, 183; also ref. 145 for earlier literature). In some cases it has even been shown that DNA extracted from bacteria after heat treatment presents single-strand breaks (15, 179). On the other hand, both in mammalian tumor cells (cf. Sect. 3) and in microorganisms (44), interruption of DNA synthesis appears to be the fastest and most severe lesion occurring upon thermal exposure.

Three different kinds of DNA damage might be expected to be caused by heat: denaturation (i.e., strand dissociation or separation), depurination, and phosphate ester-bond breakage (163). Usually, however, these lesions do not occur in the supranormal range, which may be of clinical interest, but at much higher temperatures. Although the exact temperature varies with the base composition of DNA and with buffer composition and pH, it seldom is, at least for naturally occuring DNAs, below 50°C (90, 101, 163: see also Fig. 2). Nor does it seem probable that, in heat-susceptible cells, DNA is denatures at lower temperatures as compared to DNA from heat-resistant cells, because similar melting temperatures have been found for DNAs from mesophilic and thermophilic bacteria (45, 100).

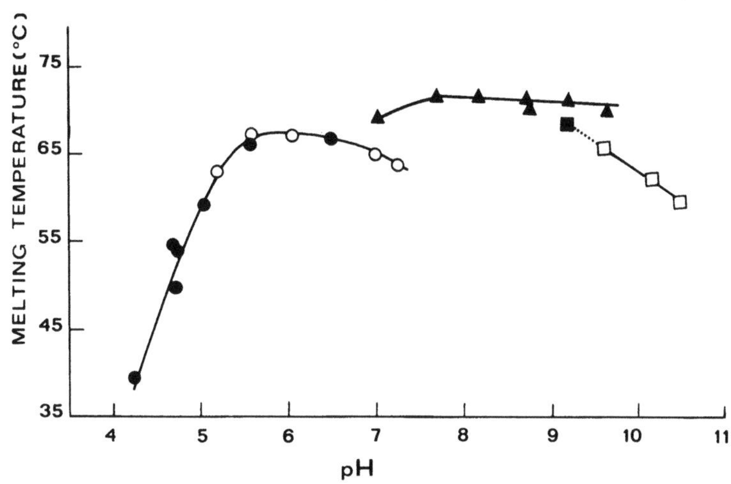

Fig. 2. Variations of melting temperature (Tm) of calfthymus DNA in various buffers at constant ionic strength 0.01: (●) Acetate; (o) phosphate; (▲) tris; (□) borate; (■) carbonate. From LEWIN and PEPPER (90)

A further argument against a direct role of such DNA lesions in the genesis of cell damage by supranormal temperatures has been amply discussed by WESTRA and DEWEY (32, 175): while ARRHENIUS plots of the temperature dependence of the cell damage give, in several systems (bacterial or yeast cells as well as mammalian normal and/or neoplastic cells), activation energies between 85 and 150 kcal/mole (79, 174, 179), the activation energies for heat-induced damages to DNA usually fall in the 25-34 kcal/mole range (40, 61); under conditions (e.g., at low pH values) that lower the melting temperature of DNA to the 40-45°C range, the enthalpy of denaturation and the free energy of stabilization of native DNA are both further decreased (128). It can therefore be said that, questionable as it may be to assume that there is a precise correlation of in vitro thermodynamic data to the in vivo situation, the comparison of the activation energies is an argument against a direct involvement of DNA in the heat-induced cell damage.

On the other hand, there is a conspicuous difference in the sensitivities of the various phases of the cell cycle to heat and to X-irradiation, the S phase being most susceptible to damage by heat (32, 127, 140, 175) but least sensitive to X-rays (8, 31).

Since X-rays are known to have a selective effect upon DNA structure (164), it is not probable that heat too affects the same target (32). If, as will be discussed later, deoxyribonucleoproteins are involved in the mechanism of cell damage by supranormal temperatures, it is therefore likely that, rather than DNA itself, the real target is either the protein component of the complex or the system, also of protein nature, which catalyzes or regulates the synthesis of the nucleic acid as well as its expression.

The possibility that supranormal temperatures may inactivate the enzyme(s) necessary for repair of previously damaged DNA is supported by the finding that, both in vivo (113, 120, 134, 180) and *in vitro* (10, 11, 63), hyperthermic exposure sensitizes tumor cells to the action of X-radiation and/or of some chemotherapeutic drugs known to have DNA as their specific target.

A quite different, and at present rather remote, hypothesis on a possible mechanism for the damage caused by supranormal temperatures stems from studies on circular DNAs. Temperature shifts either of thermosensitive DNA-replication bacterial mutants (60), or of mammalian cells transformed by a thermosensitive mutant virus (28) are able to modify the levels of catenated dimeric and oligomeric circular DNAs, either of viral or of plasmid origin. In the case of the kanamycin-resistant thermosensitive R factor Rts1, it has been found (38) that plasmid DNA exists as covalently closed circles only at permissive temperatures, whereas at nonpermissive ones, plasmid DNA is still synthesized but does not attain its final circular shape.

Also, in chick embryo fibroblasts transformed by a ts mutant of Schmidt-Ruppin-Rous sarcoma virus, the level of cateated dimeric and oligomeric mitochondrial DNA is temperature-dependent and correlates positively with the phenotypic manifestations of transformation (membrane glycoprotein pattern, deoxyglucose transport through the cell membrane): at the permissive temperature (36°C), the level of this DNA species is higher in transformed cells, and reversibly decreases to the levels of uninfected cells upon shift to the nonpermissive temperature of 41°C (115). However, since the protein synthesis inhibitor cycloheximide modifies, in a direction adverse to that occuring upon heat treatment, the levels of this catenated oligomeric mitochondrial DNA, it appears that, in this case also, the effect observed is not directly on DNA structure, but rather on some step in its biosynthesis.

b) RNA
Some reports have indicated that messenger RNA (18), and even more ribosomal RNA (136, 154), are damaged upon thermal exposure of unicellular microorganisms. The less compact structure of RNAs -- as compared to double-stranded DNA -- would allow a certain degree of conformational flexibility also at temperatures as low as 40°C (71). A real denaturation process occurs at somewhat higher temperatures; the melting temperature, which depends on the buffer composition (12, 13), is, however, generally lower than that required for DNA.

In thermophilic bacilli, ribosomal RNA has been reported to have a greater heat stability than that of *E. coli* (98, 138, 165); however, since the presence of the ribosomal protein increased the heat stability of the RNA, it is possible that ribosomal proteins are mainly responsible for the greater heat stability of the thermophilic bacilli.

As for tRNAs, whereas those from B. stearothermophilus have been reported to have the same heat stability as those from Escherichia coli (3, 52, 138), those from the extreme thermophile Thermus aquaticus appear to be more thermostable (184, 185). On the other hand, FRESCO et al. (49) have shown, using a valyl-tRNA and valyl-tRNA-synthetase of different heat stability, that the t-RNA native conformation, which can be lost on heating, is essential for synthetase recognition.

There is so far no real evidence for or against an actual involvement of some RNA species in hyperthermic cell damage. However, as has already been mentioned and as will be discussed at length below, since heat treatment appears to cause a cessation of the biosynthetic mechanism leading to RNA formation and maturation, the hypothesis that RNA itself is a target of high temperature may be superfluous -- exspecially because a lesion of a single RNA species can hardly account for all the effects of supranormal temperatures.

c) Proteins, Enzymes, and Metabolic Processes

Thermal denaturation of proteins and of enzymes is a well-known phenomenon, which has been studied in considerable detail (see, e.g., ref. 14 for theoretical treatment and for further references).

Most enzymes show, in vitro, a maximum activity at a temperature equal or near to the growth temperature of the organism from which they have been purified (i.e., 37-38°C for mammalian enzymes). The hypothesis that proteins are involved in cell damage caused by heat is substantiated by studies on thermophilic microorganisms, the proteins of which have been found to be much more resistant to heat denuration than analogous proteins of mesophilic microorganisms (3, 46, 84, 99, 181, 184, 185). Conversely, in several ts mutant strains a decreased thermal stability of some enzymes has been ascertained (39, 157).

A further indication that proteins are probably involved in heat-induced cell damage has been put forward, as already mentioned, by WESTRA and DEWEY (32, 175): the activation energy for thermal denaturation of several proteins is in the same range (100-200 kcal/mole vs. 85-150 kcal/mole) as that for cessation of cell growth and cell viability.

As mentioned, exposure to heat during the M phase seems to exert a specific denaturing action on spindle proteins (32, 175). Apart from this phase of the cell cycle, hyperthermic damage seems, both in neoplastic mammalian cells and in simpler systems, to involve mainly proteins which, either structurally or for biosynthetic purposes, are intimately connected with nucleic acids. In polyovirus I, LWOFF (96) has attributed thermal inhibition of virus multiplication to the lack of transformation of newly synthetized RNA polymerase into an enzymatically active form. Heat-inactivated viruses can be "rescued" by superinfection of the host cells with a related virus, even if the genome of the latter has been damaged by nitrogen mustard (47, 65, 80). Several temperature-sensitive mutations, in viruses as well as in microorganisms and in mammalian cell lines, are either amenable to a higher heat sensitivity of enzymes involved in biosynthetic processes, or, more often, appear to be due to increased thermal lability of an aporepressor protein, so that the induction or the repression systems of some biosynthetic pathways become heat-sensitive (54, 73, 130, 169). On the other hand it is well known that in higher organisms exposures to moderate heat, rather than causing mutations, often brings about phenotypic changes, which probably can be attributed to alterations in the processes of gene expression (24).

Even the observations of chromosomal aberrations and/or a higher frequency of mutations can sometimes be reconduced to some damage at the level of those proteins which are combined with nucleic acids in the chromosomes. It is even known that the extent of heat inactivation of transforming DNA is dramatically increased when there is protein associated with it (57, 164), possibly because of denaturation of chromosomal protein and subsequent aggregation between DNA and protein (32). Alternatively, some enzyme essential for the preservation of DNA integrity may be thermolabile (140).

A more direct involvement, as far as the effects of hyperthermia on mammalian tumor cells are concerned, of enzymes connected to DNA metabolism is demonstrated not only by the in vitro inhibitory effect of supranormal temperatures on the rate of DNA synthesis (108) but also by the finding that the sensitivity of tumor cells to DNA-specific agents (such as X-rays or certain chemotherapeutic drugs) is increased, both in vivo and in vitro, by previous or simulteneous exposure to heat (10, 11, 63, 113, 120, 134, 180). This synergism, which is of the utmost importance for its therapeutic implications, especially since it is more marked under hypoxic conditions (133), appears to be due to the inhibition, upon thermal exposure, of enzymatic process(es) necessary for the repair of sublethal damage to DNA (10, 11, 63). As for the identification of the step(s) responsible for this effect of hyperthermia, ORMEROD and STEVENSON (116) have shown that, in L5178Y murine lymphoma cells, the initial rate of rejoining of X-ray-induced single strand breaks is the same at 37°C and at 43°C, but that at higher temperature there is a concurrent DNA degradation, most probably due to the action of an exonuclease.

Other heat-induced lesions at the metabolic level concern the inhibition of RNA synthesis, especially at the level of ribosomal RNA maturation (2, 35, 70, 89, 108, 161, 162, 172), polyribosome disaggregation (70, 89, 106), and inhibition of protein synthesis (106, 108). The first of these effects, which appears to be of the utmost importance, has been described by several authors in various cellular systems. In all cases, heat adversely affects the maturation process of nucleolar RNA into ribosomal RNA (2, 70, 161, 162, 172). Upon short incubation of Ehrlich ascites cells at 44°C, for instance, uridine incorporation into nucleolar pre-RNA is irreversibly and severely decreased, the mechanism involved being that of a maturation block at the level of high-molecular weight (⩾ 41S) pre-rRNA (161, and Fig. 3). A similar, though reversible, lesion has been found to occur also in long-established "normal" cell lines (2, 70, 172).

In the ts 422E mutant of the BHK21 cell line, TONIOLO et al. have shown that the higher temperature causes a selective block of the maturation of 32S to 28S rRNA; this finding seems, however, to be restricted to that specific case; in other cell lines and in tumor cells, formation of both 28S and 18S rRNA is inhibited (161, 172).

In this more general case, WAROCQUIER and SCHERRER (172) indicate three possibilities to account for the inhibition by heat of the formation of mature ribosomal RNAs:

1. Formation, in the nucleolus, of altered 80S pre-ribosomal particles, refractory to correct maturation, as a consequence of faulty structure of either 45S pre-rRNA or of the protein associated with it in the 80S particle

2. Thermosensitivity of some enzyme required for correct maturation of the pre-rRNA

3. Thermosensitivity of some step in the process of correct methylation of pre-rRNA

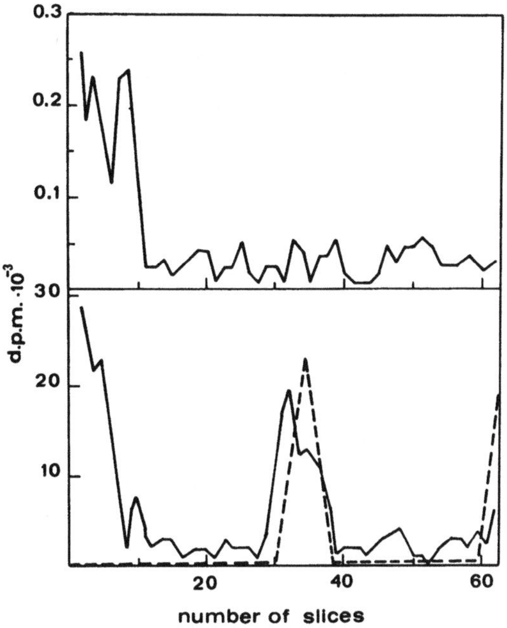

Fig. 3. Fractionation, by electrophoresis on 2.6% polyacrylamide gels ins the presence of 0.2% SDS, of nucleolar RNA from Ehrlich ascites cells preincubated 30 min at 37°C (below) or 44°C (above), then kept 45 min at 32°C in presence of 2 µCi/ml-labeled uridine.

Dashed line: distribution of ^{14}C-labeled standard of ribosomal RNA. Slices were 1.5-mm thick. From STROM et al. (161)

It should be noted that extensive methylation processes have been shown to be necessary at various stages of the synthesis of several types of RNA as well as of DNA (29, 51, 64, 73), so that a lesion at this level might be sufficient to account not only for the inhibition of RNA synthesis but also for some of the other above-mentioned effects of thermal exposure.

The prominent role of rRNA synthesis is the genesis of heat-induced cell damage is confirmed by investigations on the simultaneous action of heat and of some specific metabolic inhibitors of DNA and protein synthesis that decrease the thermal sensitivity of HeLa cells, while agents interfering with RNA synthesis act as potentiators of hyperthermia. On the other hand, McCORMICK and PENMAN (106) always found that heat-induced disaggregation of polyribosomes is reduced upon addition of actidione, and inhibitor of protein synthesis, and made instead more severe by actinomycin D, which inhibits DNA transcription into RNA. Similarly, a protective effect of the protein inhibitor cycloheximide has been shown in yeast cells by SCHENBERG-FRASCINO (139). It should, however, be remembered that heat may modify the effectiveness of these metabolic inhibitors; to assume that an increased or decreased efficiency of a given inhibitor really reflects the involvement of a given metabolic pathway in the genesis of heat-induced cell damage may therefore be often questionable. Nevertheless, these results appear to indicate that the extent of hyperthermic damage is increased when RNA synthesis acts in the opposite direction. The former result seems to confirm involvement of RNA biosynthetic processes in hyperthermia-induced cell damage, while the latter finding has been taken (106,

122, 139) as indicating that this cell damage is mediated by some newly synthesized protein. In this connection, it should be mentioned that RIEBER and BACALAO (132) have found a selective enhancement of the synthesis of a cytoplasmic protein upon heat exposure of a ts Chinese hamster cell line.

B. Membrane Systems

As an alternative to the hypothesis of a selective action of elevated temperatures on some specific macromolecules so as to bring about inhibition of some biosynthetic process, the possibility should be taken into consideration that some membrane system in the cell may be altered by heat exposure. Such a mechanism, which would account for a wide range of effects at different levels, may either involve the plasma membrane or some of the various intracellular membrane systems.

a) Plasma Membrane

The hypothesis of a primary lesion of the plasma membrane of the cell is worth special consideration. On the basis of recent experimental evidence indicating that the cell membrane is involved in the definition of neoplastic cell transformation (41, 103), is may furnish a basis for the selective thermal sensitivity of neoplastic cells.

This possibility is substantiated by the fact that several lipid membrane components -- e.g., lauroyl-1,2-diglyceride (94), 2,3-dipalmitoyl-1-phosphatidylcholine (22, 30), 2-oleyl-3-stearoyl-1-phosphatidylcholine (21), and even cholesterol (155) -- undergo sharp mesomorphic transitions in the $38-44^{o}C$ range. Also, in various complex membrane systems, both artificial and natural, it has been shown that, in this temperature range, phase transitions may occur (95, 156), sometimes also involving membrane-bound enzyme systems (75, 81, 85, 118, 144, 169). Membrane fluidity in fact plays an important role in cellular processes such as permeability (86), transport of nutrients (177), and regulation of enzyme activity (97). In several microorganisms growing at different temperatures and/or conditions, the lipid composition of the cell membrane is regulated by the growth temperature so as to attain, at this temperature, the "right" degree of fluidity (20, 27, 42, 43, 53, 74, 97, 102, 124, 151, 153).

This implies that fast variations of the growth temperature, which bring about an alteration of membrane fluidity, are likely to cause adverse effects on the cell physiology; only at longer times is an adaptation to the new temperature possible, by variation in membrane composition.

In more complex plasma membranes, the presence of cholesterol, which has the effect of decreasing the sharpness of the phase transition due to the "melting" of the hydrocarbon chains of the phospholipids (72, 87), makes, however, less probable the possibility of a sharp change of membrane fluidity upon temperature shift of a few degrees centigrade. However, since a segregation of cholesterol-free phospholipids may occur within the membrane (125), hyperthermic cell damage mediated by phospholipid phase transitions may not be excluded even in these systems, and has in fact been observed in some cases (e.g., 85). It should, moreover, be mentioned that, rather than the lipid components, heat may adversely affect also the protein components of the membrane -- similarly to what may happen within the cell (75, 83, 91, 144).

Concerning the possibility that plasma membrane is the primary site damaged by exposure to supranormal temperatures, the following results appeared to support it:

1. Temperature caused an immediate and very pronounced -- though reversible -- increase of the rate of passive diffusion of low-molecular weight substances through the membrane (162).

2. A shift from 38°C to 42°C, or even to 40°C, abruptly and irreversibly modified, in Novikoff hepatoma cells but not in their normal counterpart (i.e., in normal or regenerating liver cells) the potassium-dependent transcellular migration of glutamate (159, 160).

3. Exposure for 5 min to 47.5-48.5°C of Novikoff hepatoma cells grown in suspension culture results in an irreversible loss of the cells' capacity to transport, by a facilitated diffusion mechanism, 2-deoxyglucose (131), as well as thymidine (127) or various ribonucleotides (126). This inhibition, which from the kinetic point of view is to be attributed to an inactivation of the transport sites, without any effect on the affinity toward substrates of the remaining sites, is fairly specific the kinases required for phosphorylation of the nucleosides not being affected, and correlates well with the loss of the cells' capacity to bring about nucleic acid synthesis (126, 127).

4. Heat-induced cell damage, as measured by oxygen uptake and by thymidine incorporation into DNA, is abolished upon disruption of the neoplastic cells (108, 111).

5. An increase of the extracellular concentration of labeled uridine greatly diminishes, in intact neoplastic cells, the extent of heat-induced inhibition of uridine incorporation into RNA; similarly, in permeabilized cells the extent of nucleotide incorporation into RNA is adversely affected by heat only at relatively low concentrations of the nucleotides (162).

6. In a clonal strain of mammlian cells with increased resistance to heat shock at 46°C, REEVES (129) showed that exposure to the high temperature caused, as compared to the heat-sensitive parental line, a less marked leakage of intracellular components, particularly of uridine containing-materials.

7. Several drugs, namely the polyenic antibiotic filipin (108, 111), ethanol (108, 111), dimethylsulfoxide (142), vitamin K_3, 4-hydroxy-2-trans-octenal, and lipid hydroperoxides (4), all with a more or less marked selectivity for the lipid components of the cell membrane, show a marked synergism with thermal exposure, acting either as sensitizers for subsequent action of heat (4), or being far more effective as cell-damaging agents when acting at higher temperatures (108, 111). Although, as already mentioned, a synergism does not per se indicate a common site of action, this pattern suggested involvement of the cell membrane. On the other hand, heat-sensitive neoplastic cells appeared also to be more sensitive to some of these drugs, in particular to the polyenic antibiotics (108, 110, 111) and to prednisolone (141).

8. Preliminary sublethal heat exposure of neoplastic cells caused a significant, though slight, increase of the susceptibility of these cells to be lysed by immune antiserum + complement (Fig. 4). Conversely, an immunoresistant neoplastic cell strain appeared somewhat more heat-resistant than its immunosensitive counterpart (Fig. 5).

9. Hyperthermic exposure of neoplastic cells could, under certain conditions, increase their immunogenicity, i.e., the ability, when injected, to immunize recipient animals against subsequent transplantation of viable untreated tumor cells (23, 109).

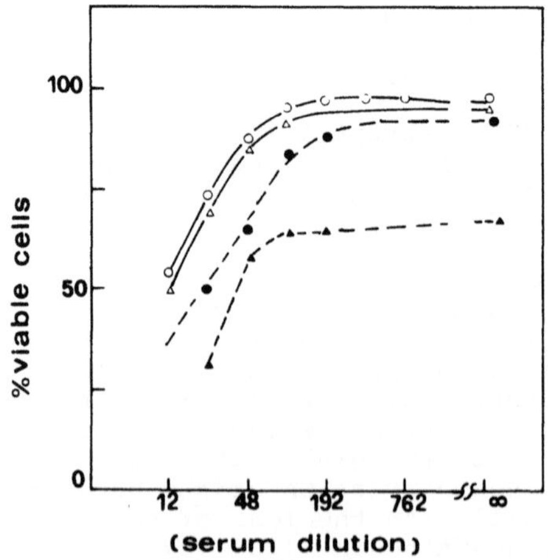

Fig. 4. Effect of previous thermal exposure on immunosensitivity of Moloney virus-induced lymphoma cells (from STROM, 158) and kept in ascites form by serial transplantation (48). Cells, obtained on 7th day after i.p. transplantation and suspended in Eagle's MEM + 10% fetal calf serum, were kept at 37.2°C (o—o, △—△) or at 42.2°C (●....●, ▲....▲) for 30 min (circles) or 60 min (triangles), then incubated for 60 min at 37°C with anti-Moloney virus antiserum, at dilution in avscissae, in presence of guinea pig complement. Cell viability was estimated by Trypan blue exclusion technique

More accurate investigations of the mechanism of heat-induced cell damage failed, however, to support this hypothesis of a primary lesion at the cell membrane level, under conditions of exposure to moderate hyperthermia (⩽ 44°C). Not only, as mentioned above, is the observed increase of passive permeability fully reversible, while the cell damage is not so, but an irreversible heat-induced alteration of membrane permeability -- such as can be revealed by loss of dye-exclusion ability -- occurs only with considerable delay after thermal exposure, (158, and Fig. 5), thus indicating that, rather than the primary lesion, it is an epiphenomenon which is revealed only upon very extensive cell damage. A similar pattern is observed by the increase of immunogenicity, which, in Ehrlich ascites cells, reached its peak only after 3 h at 42.5°C (109).

Moreover, analysis of RNA metabolism showed that the gel-electrophoresis pattern of nucleolar RNAs, which is found after thermal exposure of tumor cells, and which indicates a block in the maturation of pre-rRNA (Fig. 3), is quite different from the pattern observed in cells kept at a lower temperature but in which the permeability of the plasma membrane had been increased by addition of lucensomycin (161). On the other hand, analysis of uridine metabolites indicated that in heat-treated tumor cells, in which RNA synthesis was severely inhibited, uridine permeation in the cell was not always sensibly affected (162).

It appears therefore that, although heat treatment of tumor cells also brings about modification at the cell membrane level, these can hardly be considered as the primary effect of the thermal exposure, but are rather the consequence of previous cellular lesions.

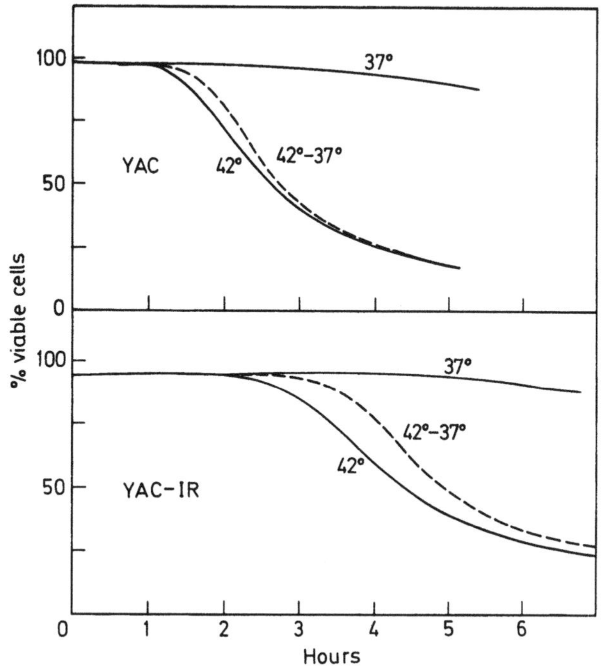

Fig. 5. Effect of elevated temperature on cell viability of Moloney virus lymphoma cells (from STROM, 58).
YAC was same lymphoma mentioned in Fig. 4. YAC-IR is a variant of YAC, which exhibited in vitro immunoresistance (48). Cells, suspended in Eagle's MEM + 10% fetal calf serum, were incubated at 37°C or at 42°C, or kept 40 min at 42°C and then transferred to 37°C. Cell viability was estimated by Trypan blue exclusion technique

b) Mitochondria, Lysosomes, and Other Intracellular Membrane Systems

Of the various membrane systems and/or organelles present within the cell, mitochondria do not appear as a possible relevant target, since they maintain their morphological integrity (70, 89), while respiration remains unaffected during the first stages of thermal exposure (see e.g. Fig. 1).

As for the intracellular membranes, a possible role of the nuclear membrane has been hypothesized in yeast by SCHENBERG-FRASCINO and MOUSTACCHI (140) to explain the cell cycle variations of the lethal and mutagenic effects of elevated temperatures. No evidence in favor of this hypothesis has been found. Similarly, there seems to be no reason to suspect the involvement of most of the other intracellular organelles. Only for lysosomes does the possibility of a primary involvement seem feasible, since through a release of their acid hydrolases they could bring about damage to the various components of the cells, possibly even to chromosemes (1).

As mentioned above, DNA degradation has been reported to occur at 43°C in L5178Y murine lymphoma cells (116). In Novikoff hepatoma, however, although isolated lysosomes were found to be somewhat more thermolabile than those from normal or regenerating rat liver, no indication could be found that, within the cell, supranormal temperatures brought about a significant release of their hydrolytic enzymes (168). From the morphologic point of view, on the other hand, only upon extensive thermal treatment has a moderate release of lysosomal enzymes been found (119), while at earlier times these organelles appear normal, and only occasionally increased in number (70, 89).

In conclusion, the present status of knowledge on the mechanism by which thermal exposure causes a selective damage to tumor cells remains to be elucidated. Experimental evidence points, however, toward a direct involvement of some enzyme systems necessary for the biosynthesis and for the maturation of nucleic acids, especially of pre-rRNA.

REFERENCES

1. ALLISON, A.C., PATON, G.R.: Chromosome damage in human diploid cells following activation of lysosomal enzymes. Nature (Lond.) 207, 1170 (1965)
2. AMALRIC, F., SIMARD, R., ZALTA, J.P.: Effect de la temperature supra-optimale sur les ribonucléoprotéines et le RNA nucléolaire; II. Etude biochimique. Exp. Cell Res. 55, 370 (1969)
3. ARCA, M., CALVORI, C., FRONTALI, L., TECCE, G.: The enzymic synthesis of aminoacyl derivatives of soluble ribonucleic acid from Bacillus stearothermophilus. Biochim. biophys. Acta 87, 440 (1964)
4. von ARDENNE, M.: Betrachtungen und Messungen zur selektiven chemischen Sensibilisierung der Krebszellen gegen Wärme. In: Theoretische und experimentelle Grundlagen der Krebs-Mehrschritt Therapie. Von Ardenne, M. (ed.) Berlin: VEB Verlag Volk und Gesundheit, p. 298
5. von ARDENNE, M., KRÜGER, W.: Über die Auffindung eines Bereiches hoher Selektivität zwischen Krebszellen und gesunden Zellen bei der Krebstherapie mit Extremhyperthermie. Naturwissenschaften 53, 436 (1966)
6. von ARDENNE, M., REITNAUER, P.G., RIEGER, F.: Zur Erkundung therapeutischer Angriffspunkte im intermediären Stoffwechsel der Krebszelle. In: Theoretische und experimentelle Grundlagen der Krebs-Mehrschritt-Therapie. Von Ardenne, M. (ed.) Berlin: VEB Verlag Volk und Gesundheit, 1967, p. 286
7. AUERSPERG, N.: Differential heat sensitivity of cells in tissue culture, Nature (Lond.) 209, 415 (1966)
8. BACCHETTI, S., CASSANDRO, M., MAURO, F.: Radiosensitivity in relation to the cell cycle and recovery from X-ray sublethal damage in diploid yeast. Exp. Cell Res. 46, 292 (1967)
9. BENDER, E., SCHRAMM, T.: Untersuchungen zur Thermosensibilität von Tumor und Normalzellen in Vitro. Acta biol. med. germ. 17, 527 (1966)
10. BEN-HUR, E., BRONK, B.V., ELKIND, M.M.: Thermally enhanced radiosensitivity of cultures chinese hamster cells. Natur (New Biol.) 238, 209 (1972)
11. BEN-HUR, E., ELKIND, M.M., BRONK, B.V.: Thermally enhanced radioresponse of cultured chinese hamster cells. Inhibition or repair of sub-lethal damage and enhancement of lethal damage. Radiat. Res. 58, 38 (1974)
12. BOEDTKER, H.: Configurational proteries of Tobacco Mosaic Virus ribonucleic acid. J. molec. Biol. 2, 171 (1960)
13. BOGDANOVA, E.S., GAVRILOVA, L.P., DVORKIN, G.A., KISELEV, N.A., SPIRIN, A.S.: Studies on the macromolecular structure of high polymer (ribosomal) ribonucleic acid from Escherichia coli. Biokhimiya 27, 387 (1962)

14. BRANDTS, J.F.: Heat effects on proteins and enzymes. In: Thermobiology. Rose, A.H. (ed.) London: Academic Press, p. 24, 1967
15. BRIDGES, B.A., ASHWOOD-SMITH, M.J., MUNSON, R.J.: Correlation of bacterial sensitivites to ionizing radiation and mild heating. J. gen. Microbiol. 58, 115 (1969)
16. BUCCIANTE, L.: Ulteriori ricerche sulla velocità della mitosi nelle cellule coltivate in vitro in funzione della temperatura. Arch. exp. Zellforsch. 5, 1 (1928)
17. BUSCH, W.: Über den Einfluss, welchen heftigere Erysipeln zuweilen auf organisierte Neubildungen ausüben: Verhandl. Naturhist. Preuss. Rhein. Westphal. 23, 28 (1866)
18. BYFIELD, J.E., SCHERBAUM, K.: Suppression of RNA and protein accumulation by temperature shifts in a heat-synchronized protozoa. Life Sci. 5, 2263 (1966)
19. CAVALIERE, R., CIOCATTO, E.C., GIOVANELLA, B.C., HEIDELBERGER, C., JOHNSON, R.O., MARGOTTINI, M., MONDOVI, B., MORICCA, G., ROSSI-FANELLI, A.: Selective heat sensitivity of cancer cells. Cancer (Philad.) 20, 1351 (1967)
20. CHANG, S.B., MATSON, R.S.: Membrane stability (thermal) and nature of fatty acids in yeast cells. Biochem. biophys. Res. Comm. 46, 1529 (1972)
21. CHAPMAN, D.: The effect of heat on membranes and membrane constituent. In: Thermobiology. Rose, A.H. (ed.), London: Academic Press, 1967, p. 122
22. CHAPMAN, D. WILLIAMS, R.M., LADBROOKE, B.D.; Physical studies of some phospholipids. VI. Thermotropic and lyotropic mesomorphism of some 1,2-diacylphosphatidylcholines (lecithins). Chem Phys. Lipids 1, 445 (1967)
23. CHECK, J.H., CHILDS, T.C., BRADY, L.W., DERASSE, A.R., FUSCALDO, K.E.: Protection against spontaneous mouse mammary adenocarcinoma by inoculation of heat-treated syngeneic mammary tumor cells. Int. J. Cancer 7, 403 (1971)
24. CLARKE, K.U.: Insects and temperature. In: Thermobiology. Rose, A.H. (ed.), London: Academic Press, 1967, p. 293
25. CLOWES, G.H.A.: A study of the influence exerted by a variety of physical and chemical forces on the virulence of carcinoma in mice, and of the conditions under which immunity against cancer may be experimentally induced in these animals. Brit. med. J. 2, 1548 (1906)
26. COLLINS, P., LACY, D.: Studies on the structure and function of the mammalian testis. II. Cytological and histochemical observation on the testis of the rat after a single exposure to heat applied for different lengths of time. Proc. roy. Soc. B 172, 17 (1969)
27. CULLEN, J., PHILLIPS, M.C., SHIPLEY, G.G.: The effects of temperature on the composition and physical properties of the lipids of Pseudomonas fluorescens. Biochem. J. 125, 733 (1971)
28. CUZIN, F., VOGT, M., DIECKMANN, M., BERG, P.: Induction of virus multiplication in 3T3 cells tranformed by a thermosensitive mutant of polyoma virus. II. Formation of oligomeric polyoma DNA molecules. J. molec. Biol. 47, 317 (1970)
29. DAVIDSON, J.N.: The Biochemistry of the Nucleic Acids. London: Methuen and Co., 1969
30. Van DEENEN, L.L.M.: Membrane lipids. In: Regulatory Functions of Biological Membranes. Järnefelt, J., (ed.), Amsterdam: Elsevier Publ. Co., 1968, Vol. XI, p. 72
31. DEWEY, W.C., FURMAN, S.C., MILLER, H.H.: Comparison of lethality and chromosomal damage induced by X-rays in synchronized Chinese hamster cells in vitro. Radiat. Res. 43, 561 (1970)
32. DEWEY, W.C., WESTRA, A., MILLER, H.H., NAGASAWA, H.: Heat-induced lethality and chromosomal damage in synchronized Chinese hamster cells treated with 5-bromodeoxyuridine. Int. J. Radiat. Biol 20, 505 (1971)

33. DICKSON, J.A., ELLIS, H.A.: Stimulation of tumour cell dissemination by raised temperature (42°C) in rats with transplanted Yoshida tumours. Nature (Lond.) 248, 354 (1974)
34. DICKSON, J.A., MUCKLE, D.S.: Total body hyperthermia versus primary tumor hyperthermia in the treatment of the rabbit VX-2 carcinoma. Cancer Res. 32, 1916 (1972)
35. DICKSON, J.A., SHAH, D.M.: The effect of hyperthermia (42°C) on the biochemistry and growth of a malignant cell line. Europ. J. Cancer, 8, 561 (1972)
36. DICKSON, J.A., SUZANGAR, M.: In vitro sensitivity screening system for human cancers to drugs and hyperthermia. Brit. J. Cancer 28, 81 (1973)
37. DICKSON, J.A., SUZANGAR, M.: In vitro-in vivo studies on the susceptibility of the solid Yoshida sarcoma to drugs and hyperthermia (42°C). Cancer Res. 34, 1263 (1974)
38. DI JOSEPH, C.G., KAJI, A.: The thermosensitive lesion in the replication of the drug resistance factor, Rts1. Proc. nat. Acad. Sci. (Wash.) 71, 2515 (1974)
39. EIDLIC, L., MIEDHARDT, F.C: The thermosensitive critical event of the viral cycle. J. Bact. 89, 706 (1965)
40. EIGNER, J., BOEDTKER, H., MICHAELS, G.: The thermal degradation of nucleic acids. Biochim. biophys. Acta 51, 165 (1961)
41. EMMELOT, P., BENEDETTI, E.L.: On the possible involvement of the plasma membranes in the carcinogenic process. In: Carcinogenesis: a Broad Critique, Baltimore: Williams and Wilkins, p. 471
42. ESFAHANI, M., BARNES, E.M., WAKIL, S.J.: Control of fatty acid composition in phospholipids of Escherichia coli: response to fatty acid supplements in a fatty acid auxotroph. Proc. nat. Acad. Sci. (Wash.) 64, 1057 (1969)
43. ESSER, A.F., SOUZA, K.A.: Correlation between thermal death and membrane fluidity in Bacillus stearothermophilus. Proc. Nat. Acad. Sci. (Wash.) 71, 4111 (1974)
44. EVERSON, D.P., PRESCOTT, D.M.: Disruption of DNA synthesis in Euplotes by heat shock. Exp. Cell Res. 63, 245 (1970)
45. EVREINOVA, T.N., DAVYDOVA, I.M., SUKOVER, A.P., GORYUNOVA, S.V.: Nucleic acids of thermophilic blue-green algae Mastigocladus laminosus. Dokl. Akad. Nauk. U.S.S.R. 137, 213,(1961)
46. FARREL, J., ROSE, A.H.: Temperature effects on microorganisms. In: Thermobiology. Rose, A.H. (ed.), London: Academic Press, 1967, p. 146
47. FENNER, F., HOLMES, I.H., JOKLIK, W.K., WOODROOFE, G.M.: Reactivation of heat-inactivated pox viruses: a general phenomenon which includes the fibroma-mixoma virus transformation of Berry and Dedrick. Nature (Lond.) 183, 1340 (1959)
48. FENYÖ, E.M., BIBERFELD, P., KLEIN, E.: Studies on the relations between virus release and cellular immunosensitivity on Moloney lymphomas. J. nat. Cancer Inst. 42, 837 (1969)
49. FRESCO, J.R., ADAMS, A., ASCIONE, R., HENLEY, D., LINDAHL, T.: Tertiary structure in transfer ribonucleic acids. Cold Spr. Harb. Symp. quant. Biol. 31, 527 (1966)
50. FRIEDGOOD, H.B.: On the thermal death point of sarcoma and normal mononuclear cells (Walker rat tumor Nr. 1). Arch. exp. Zellforsch. 7, 243 (1928)
51. FRIEDKIN, M.: Enzymatic aspects of folic acid. Ann. Rev. Biochem. 32, 185 (1963)
52. FRIEDMAN, S.M., WEINSTEIN, I.B.: Protein synthesis in a subcellular system from Bacillus stearothermophilus. Biochim. biophys. Acta 114, 593 (1966)
53. FULCO, A.J.: The biosynthesis of unsaturated fatty acids by bacilli, IV. Temperature-mediated control mechanism. J. biol. Chem. 247, 3511 (1972)

54. GALLANT, J., STAPLETON, R.: Properties of a temperature-sensitive regulatory system. Proc. nat. Acad. Sci. (Wash.) 50, 348 (1963)
55. GERICKE, D.: Studies on the thermosensitivity of experimental tumors in small animals. Abstr. Tenth Internat. Cancer Congr., p. 572, Houston (1970)
56. GERICKE, D., CHANDRA, P., ORII, H., WACKER, A.: In vitro thermosensibility of experimental tumors in small animals. Naturwissenschaften 3, 135 (1971)
57. GINOZA, W., GUILD, W.R.: On the inactivation of transforming DNA by temperature below the melting point. Proc. nat. Acad. Sci. (Wash.) 47, 633 (1961)
58. GIOVANELLA, B.C., LOHMAN, W.A., HEIDELBERGER, C.: Effects of elevated temperatures and drugs on the viability of L1210 leukemia cells. Cancer Res., 30, 1623 (1970)
59. GIOVANELLA, B.C., MORGAN, A.C., STEHLIN, J.S., WILLIAMS, L.J.: Selective lethal effect of supranormal temperatures on mouse sarcoma cells. Cancer Res. 33, 2568 (1973)
60. GOEBEL, W.: Studies on extrachromosomal DNA elements. Europ. J. Biochem. 15, 311 (1970)
61. GREEN, S., ZAMENHOF, S.: Studies on depurination of DNA by heat. J. molec. Biol. 4, 123 (1962)
62. HAALAND, M.: Contributions to the study of the development of sarcomas under experimental conditions. A.R. imp. Cancer Res. Fd. p. 175 (1908)
63. HAHN, G.M.: Metabolic aspects of the role of hyperthermia in mammalian cell inactivation and their possible relevance to cancer treatment. Cancer Res., 34, 3117 (1974)
64. HALL, R.H.: The modified Nucleosides in Nucleic Acids. New York: Columbia University Press, 1971
65. HANAFUSA, T., HANAFUSA, H., KAMAHORA, J.: Transformation of Ectromelia into Vaccinia virus in tissue culture. Virology 8, 525 (1959)
66. HARRIS, M.: Criteria of viability in heat-treated cells. Exp. Cell Res. 44, 658 (1966)
67. HARRIS, M.: Temperature-resistant variants in clonal populations of pig kidney. Exp. Cell Res. 46, 301 (1967)
68. HARRIS, M.: Maturation rates at different ploidy level. J. cell. Physiol. 78, 177 (1971)
69. HARRIS, J.R., WHITMORE, G.F: Chinese hamster cells exhibiting a temperature dependent alteration in purine transport. J. cell. Physiol. 83, 43 (1974)
70. HEINE, U., SVERAK, L., KONDRATICK, J., BONAR, R.A.: The behavior of HeLa-S_3 cells under the influence of supranormal temperatures. J. Ultrastruct. Res. 34, 375 (1971)
71. HILBERS, C.W., SHULMAN, R.G., YAMANE, T., STEITZ, J.A.: High resolution proton NMR study of an isolated fragment of R17 bacteriophage in RNA. Nature (Lond.) 248, 225 (1974)
72. HINZ, H.J., STURTEVANT, J.M.: Calorimetric investigation of the influence of cholesterol on the transmission properties of bilayers formed from synthetic L-α-lecithins in aqueous suspension. J. biol. Chem. 247, 3697 (1972)
73. HORIUCHI, T., NOVICK, A.: A thermolabile repression system. Cold Spr. Harb. Symp. quant. Biol. 26, 247 (1961)
74. HUANG, L., LORCH, S.K., SMITH, G.G., HUANG, A.: Control of membrane lipid fluidity in Acholeplasma laidlawii. FEBS Lett. 43, 1 (1974)
75. JACKSON, W.M., KOSTYLA, J., NORDIN, J.H., BRANDTS, J.F.: Calorimetric study of protein transitions in human erythrocyte ghosts. Biochemistry 12, 3662 (1973)
76. JARES, J.J., WARREN, S.L.: Physiological effects of radiation. I. A study of the in vitro effect of high fever temperatures upon certain experimental animal tumors. Amer. J. Roentgenol. 41, 685 (1939)

77. JENSEN, C.O.: Experimentelle Untersuchungen über Krebs bei Mäusen; Zbl. Bakt. 34, 28 (1903)
78. JOHNSON, H.J.: The action of short radio waves on tissues. III. A comparison of the thermal sensitivities of transplantable tumors in vivo and in vitro. Amer. J. Cancer 38, 533 (1940)
79. JOHNSON, F.H., EYRING, H., PALISSAR, M.J.: The Kinetic Basis of Molecular Biology. New York: J. Wiley and Sons, 1954
80. JOKLIK, W.K., ABEL, P., HOLMES, I.H.: Reactivation of pox viruses by an non-genetic mechanism. Nature (Lond.) 186, 992 (1960)
81. KABACK, H.R.: The role of the phosphoenolpyruvate-phosphotransferase system in the transport of sugars by isolated membrane preparations of Escherichia coli. J. biol. Chem. 243, 3711 (1968)
82. KACHANI, Z.F.C., SABIN, A.B.: Reproductive capacity and viability at higher temperatures of various transformed hamster cell lines. J. nat. Cancer Inst. 43, 469 (1969)
83. KELLY, L.E.: Temperature-sensitive mutations affecting the regenerative sodium channel in Drosophila melanogaster. Nature (Lond.) 248, 166 (1974)
84. KOFFLER, H., GALE, G.O.: The relative thermostability of cytoplasmic proteins from thermophili bacteria. Arch. Biochem. 67, 249 (1957)
85. de KRUYFF, B., van DIJCK, P.W.M., GOLDBACH, R.W., DEMEL, R.A., van DEENEN, L.L.M.: Influence of fatty acid and sterol composition on the lipid phase transition and activity of membrane-bound enzymes in Acholeplasma laidlawii. Biochim. biophys. Acta 330, 269 (1973)
86. de KRUYFF, B., De GREEF, W.J., Van EYK, R.V.W., DEMEL, R.A., van DEENEN, L.L.M.: The effect of different fatty acid and sterol composition on the erythritol flux through the cell membrane of Acholeplasma laidlawii. Biochim. biophys. Acta 298, 479 (1973)
87. LADBROOKE, B.D., WILLIAMS, R.M., CHAPMAN, D.: Studies on lecithin-cholesterol-water interactions by differential scanning calórimetry and X-ray diffraction. Biochim. biophys. Acta 150, 333 (1968)
88. LAMBERT, R.A.: Demonstration of the greater susceptibility to heat of sarcoma cells as compared with actively proliferating connective-tissue cells. J. Amer. med. Ass. 59, 2147 (1912)
89. LEVINE, E.M., ROBBINS, E.B.: Differential temperature sensitivity of normal and cancer cells in culture. J. cell. Physiol. 76, 373 (1970)
90. LEWIN, S., PEPPER, D.S.: Variation of the melting temperature of calf-thymus DNA with pH and type of buffer. Arch. Biochem. 109, 192 (1965)
91. LIN, P.S., WALLACH, D.F.H., TSAI, S.: Temperature-induced variations in the surface topology of cultured lymphocytes are revealed by scanning electron microscopy. Proc. nat. Acad. Sci. (Wash.) 70, 2492 (1973)
92. LOEB, L.: Über Transplantation von Tumoren. Virchows Arch. path. Anat. 172, 345 (1903)
93. LOVE, R., SORIANO, R.Z., WALSH, R.J.: Effect of hyperthermia on normal and neoplastic cells in vitro. Cancer Res. 30, 1525 (1970)
94. LUTTON, E.S.: Thermal properties and crystal state. In: Fatty Acids. Markley, K.S. (ed.) New York: Interscience Publ. 1967, Pt. 4, p. 2583
95. LUZZATI, V., HUSSON, F.: The structure of the liquid-crystalline phases of lipid-water systems. J. Cell Biol. 12, 207 (1962)
96. LWOFF, A.: The thermosensitive critical event of the viral cycle. Cold Spr. Harb. Symp. quant. Biol. 27, 159 (1962)
97. LYONS, J.M.: Phase transitions and control of cellular metabolism at low temperatures. Cryobiology 9, 341 (1972)

98. MANGIANTINI, M.T., TECCE, G., TOSCHI, G.: Molecular properties and thermal stability of ribonucleic acid in a thermophilic microorganism. Suppl. Nuovo Cimento 25, 45 (1962)
99. MANNING, G.B., CAMPELL, L.L.: Thermostable α-amylase of Bacillus Stearothermophilus. J. biol Chem. 236, 2952 (1961)
100. MARMUR, J.: Thermal denaturation of deoxyribosenucleic acid isolated from a thermophile. Biochim. biophys. Acta 38, 342 (1960)
101. MARMUR, J., DOTY, P.: Heterogeneity in deoxyribonucleic acids. I. Dependence on composition of the configurational stability of deoxyribonucleic acids. Nature (Lond.) 183, 1427 (1959)
102. MARR, A.G., INGRAHAM, J.L.: Effect of temperature on the composition of fatty acids in Escherichia coli. J. Bact. 84, 1260 (1962)
103. MARX, J.L.: Biochemistry of cancer cells: focus on the cell surface. Science 183, 1279 (1974)
104. MARTIN, R.J., SCHLOERB, P.R.: Induction of mitotic synchrony by intermittent hyperthermia in the Walker 256 rat carcinoma. Cancer Res. 24, 1997 (1964)
105. McBURNEY, M.W., WHITMORE, G.F.: Selection for temperature-sensitive mutants of diploid and tetraploid mammalian cells. J. cell. Physiol. 83, 69 (1974)
106. McCORMICK, W., PENMAN, S.: Regulation of protein syntheis in HeLa cells: translation at elevated temperatures. J. molec. Biol. 39, 315 (1969)
107. MEISS, H.K., BASILICO, C.: Temperature sensitive mutants of BHK 21 cells. Nature (New Biol.) 239, 66 (1972)
108. MONDOVI, B., FINAZZI-AGRO, A., ROTILIO, G., STROM, R., MORICCA, G., ROSSI-FANELLI, A.: The biochemical mechanism of selective heat sensitivity of cancer cells. II. Studies on nucleic acids and protein synthesis. Europ. J. Cancer 5, 137 (1969)
109. MONDOVI, B., SCIOSCIA SANTORO, A., STROM, R., FAIOLA, R., ROSSI-FANELLI, A.: Increased immunogenicity of Ehrlich ascites cells after heat treatment. Cancer (Philad.) 30, 885 (1972)
110. MONDOVI, B., STROM, R., FINAZZI-AGRO, A., CAIAFA, P., DE SOLE, P., BOZZI, A., ROTILIO, G., ROSSI-FANELLI, A.: Effects of polyenic antibiotics on Ehrlich ascites and Novikoff hepatoma cells. Cancer Res. 31, 505 (1971)
111. MONDOVI, B., STROM, R., ROTILIO, G., FINAZZI-AGRO, A., CAVALIERE, R., ROSSI-FANELLI, A.: The biochemical mechanism of selective heat sensitivity of cancer cells. I. Studies on cellular respiration. Europ. J. Cancer 5, 129 (1969)
112. MUCKLE, D.S., DICKSON, J.A.: The selective inhibition effect of hyperthermia on the metabolism on growth of malignant cells. Brit. J. Cancer 25, 771 (1971)
113. MUCKLE, D.S., DICKSON, J.A.: Hyperthermia (42°C) as an adjuvant to radiotherapy and chemotherapy in the treatment of the allogeneic VX-2 carcinoma in the rabbit. Brit. J. Cancer 27, 307 (1973)
114. MULLER, H.J.: The measurement of gene mutation rate in Drosophila, its high variability and its dependence upon temperature. Genetics 13, 279 (1928)
115. NASS, M.M.K.: Temperature-dependent formation of dimers and oligomers of mitochondrial DNA in cells transformed by an thermosensitive mutant of Rous sarcoma virus. Proc. nat. Acad. Sci. (Wash.) 70, 3739 (1973)
116. ORMEROD, M.G., STEVENS, U.: The rejoining of X-ray-induced strand breaks in the DNA of a murine lymphoma cell (L5178Y). Biochim. biophys. Acta 232, 72 (1971)
117. OSSOVSKI, L., SACHS, L.: Temperature sensitivity of polyoma virus, induction of cellular DNA synthesis, and multiplication of transformed cells at high temperature. Proc. nat. Acad. Sci. (Wash.) 58, 1938 (1967)

118. OVERATH, P., SCHAIRER, H.U., STOFFEL, W.: Correlation of in vivo and in vitro phase transitions of membrane lipids in Escherichia coli. Proc. nat. Acad. Sci (Wash.) 67, 606 (1970)
119. OVERGAARD, K., OVERGAARD, J.: Investigation on the possibility of a thermic tumor therapy. I. Short wave treatment of a transplanted isologous mouse mammary carcinoma. Cancer (Philad.) 8, 65, (1972a)
120. OVERGAARD, K., OVERGAARD, J.: Investigations on the possibility of a thermic tumor therapy. II. Action of combined heat-Roentgen treatment on a transplanted mouse mammary carcinoma. Europ. J. Cancer 8, 572 (1972b)
121. PALZER, R.J., HEIDELBERGER, C.: Studies on the quantitative biology of hyperthermic killing of HeLa cells. Cancer Res. 33, 415 (1973a)
122. PALZER, R.J., HEIDELBERGER, C.: Influence of drugs and synchrony on the hyperthermic killing of HeLa cells. Cancer Res. 33, 422 (1973b)
123. PETRAKIS, N.L., LIEBERMAN, E.: The thermal susceptibility of leucocytes from normal and leukemic subjects. Clin. Res. Proc. 4, 53 (1956)
124. PEYPOUX, F., MICHEL, G.: Etude des phospholipides d'une souche sauvage et d'une souche mutante thermosensible d'Escherichia coli. Biochim. biophys. Acta 218, 453 (1970)
125. PHILLIPS, M.C.: Physical state of phospholipids and cholesterol in monolayers, bilayers and membranes. In: Progress in Surface and Membrane Science. Danielli, J.F., Rosenberg, M.D., Cadenhead, D.A. (eds.). New York: Academic Press, 1972, Vol. V, p. 139
126. PLAGEMANN, P.G.W.: Nucleoside transport by Novikoff rat hepatoma cells. Specificity and mechanism of transport. Reactions and relationship to nucleoside incorporation into nucleic acids. Biochim. biophys. Acta. 233, 688 (1971)
127. PLAGEMANN, P.G.W., ERBE, J.: Thymidine transport by cultured Novikoff hepatoma cells and uptake by simple diffusion and relationship to incorporation into deoxyribonucleic acid. J. Cell Biol. 55, 161 (1972)
128. PRIVALOV, P.L.: Thermal investigation of biopolymer solutions and scanning microcalorimetry. FEBS Lett. 40, S 140 (1974)
129. REEVES, O.R.: Mechanism of acquired resistence to acute heat shock in cultured mammalian cells. J. cell. Physiol. 79, 157 (1972)
130. RENGER, H.C., BASILICO, C.: Mutation causing temperature-sensitive expression of cell transformation by tumor virus. Proc. nat. Acad. Sci. (Wash.) 69, 109 (1972)
131. RENNER, E.D., PLAGEMANN, P.G.W., BERNLOHR, R.W.: Permeation of glucose by simple and facilitated diffusion by Novikoff rat hepatoma cells in suspension culture and its relationship to glucose metabolism. J. biol. Chem. 247, 5765 (1972)
132. RIEBER, M., BACOLAO, J.: Selective enhancement of cytoplasmic protein species and continued histone synthesis during inhibition of DNA synthesis in temperature-sensitive Chinese hamster cells. Cancer Res. 34, 3083 (1974)
133. ROBINSON, J.E., WIZENBERG, M.J., McCREADY, W.A.: Combined hyperthermia and radiation suggest an alternative to heavy particle therapy for reduced oxygen enhacement ratios. Nature (Lond.) 251, 521 (1974)
134. ROBINSON, J.E., WIZENBERG, M.J., McCREADY, W.A., SCHELTEMA, J.: Thermally enhanced radiocurability of mammary tumor in C3H Mice. Proc. IV Intern. Biophys. Congress, Vol. 1, comm. EIII d 3/11, p. 289, Moscow (1972)
135. ROHDENBURG, G.L., PRIME, F.: Effect of combined radiation and heat on neoplasms. Arch. Surg. (Chicago) 2, 116 (1921)
136. ROSENTHAL, L.J. JANDOLO, J.J.: Thermally induced intracellular alterations of ribosomal ribonucleic acid. J. Bact. 103, 833 (1970)

137. SAPOZINSK, M.D., DESCHNER, E.E., HAHN, E.W.: Induction of mitotic synchrony by intermittent hyperthermia in Ehrlich carcinoma in vivo. Nature (Lond.) 244, 299 (1973)
138. SAUNDERS, F.G., CAMPELL, L.L.: Ribonucleic acid and ribosomes of Bacillus stearothermophilus. J. Bact. 91, 332 (1966)
139. SCHENBERG-FRASCINO, A.: Lethal and mutagenic effects of elevated temperature on haploid yeast. II. Recovery from thermolesions. Molec. gen. Genet. 117, 239 (1972)
140. SCHENBERG-FRASCINO, A., MOUSTACCHI, E.: Lethal and mutagenic effects of elevated temperature on haploid yeast. II. Recovery from thermolesions. Molec. gen. Genet. 115, 243 (1972)
141. SCHRECK, R.: Prednisolone-sensitivity and cytology of viable lymphocytes as tests for chronic lymphocytic leukemia. J. nat. Cancer Inst. 33, 837 (1964)
142. SCHRECK, R.: Effect of heat and dimethyl sulfoxide on normal and leukemic lymphocytes. Lancet (1965) ii, 1020
143. SCHREK, R.: Sensitivity of normal and leukemic lymphocytes and leukemia myeloblasts to heat. J. nat. Cancer Inst. 37, 649 (1966)
144. SCHWENCKE, J. MAGANA-SCHWENCKE, N.: Temperature effects on the proline transport system of Saccaromyces chevalieri. Biochim. biophys. Acta 241, 513 (1971)
145. SELAWRY, O.S., GOLDSTEIN, M.M., McCORMICK, T.: Hyperthermia in tissue-cultured cells of malignant origin. Cancer Res. 17, 785 (1957)
146. SHERMAN, F.: A study of the effects of elevated temperatures on the growth and inheritance of Saccaromyces cervisiae. Ph. D. thesis, University of California, UCRL - 8573 (1958)
147. SIMARD, R.: Une fonction cellulaire thermosensible reversible: la synthèse du RNA nucléolaire. J. Micr. 6, 82a (1967)
148. SIMARD, R., AMALRIC, F., ZALTA, J.P.: Effect de la température supra-optimale sur les ribonucléoprotéines et le RNA nucléolaire; I. Etude ultrastructurale. Exp. Cell. Res. 55, 359 (1969)
149. SIMARD, R., BERNHARD, W.: A heat-sensitive cellular function located in the nucleolus. J. Cell Biol. 34, 61 (1967)
150. SIMARD, R., LANGELIER, Y., MANDEVILLE, R., MAESTRACCI, N., ROYAL, A.: Inhibitors as tools in elucidating the structure and the function of the nucleus. In: The Cell Nucleus. Busch, H. (ed.), Vol. 3, 447. New York: Academic Press (1974)
151. SINENSKY, M.: Homeoviscous adaption. A homeostatic process that regulates the viscosity of membrane lipids in Escherichia coli. Proc. nat. Acad. Sci. (Wash.) 71, 522 (1974)
152. SISKEN, J.E., MORASCA, L., KIBBY, S.: Effects of temperature on the kinetics of the mitotic cycle of mammalian cells in culture. Exp. Cell Res. 39, 103 (1965)
153. SMITH, G.G., RUWART, M.T., HUANG, A.: Lipid phase transitions in membrance vescicles from Thermoplasma acidophila. FEBS Lett. 45, 96 (1974)
154. SOGIN, S.J., ORDAL, Z.J.: Regeneration of ribosomes and ribosomal ribonucleic acid during repair of thermal injury to Staphylococcus aureus. J. Bact. 94, 1082 (1967)
155. SPIER, H.L., van SENDEN, K.G.: Phase transition of cholesterol. Steroids 6, 871 (1965)
156. STEIM, J.M., TOURTELLOTTE, M.E., REINERT, J.C., McELHANEY, R.N., RADER, R.L.: Calorimetric evidence for the liquid crystalline state of lipids in a biomembrane. Proc. nat. Acad. Sci. (Wash.) 63, 104 (1969)
157. STREISINGER, G., MUKAI, F., DREYER, W.J., MILLER, B., HORIUCHI, S.: Mutations affecting the lysozyme of phage T4. Cold Spr. Harb. Symp. quant. Biol. 26, 25 (1961)
158. STROM, R.: Ricerche sul meccanismo d'azione del calore sui tumori. Atti Soc. It. Cancerol. Turin 1970, Vol. VII, Pt. 2, p. 49

159. STROM, R., CAIAFA, P., MONDOVI, B., ROSSI-FANELLI, A.: Effect of temperature on potassium-dependent stimulation of transcellular migration in normal and neoplastic cells. FEBS Lett. 3, 343 (1969)
160. STROM, R., CAIAFA, P., MONDOVI, B., ROSSI-FANELLI, A.: Temperature dependence of transcellular migration of glutamate in normal and neoplastic cells. In: Permeability and function of Biological Membrances. Bolis, L., Katchalsky, A., Keynes, R.D., Lowenstein, W.R., Pethica, B.A. (eds.). Amsterdam: North Holland Publ. Co., 1970, p. 333
161. STROM, R., CRIFO, C., BOZZI, A., ROSSI-FANELLI, A.: Inhibition by elevated temperatures of ribosomal RNA maturation in Ehrlich ascites cells. Cancer Biochem. Biophys. 1, 57 (1975)
162. STROM, R., SCIOSCIA SANTORO, A., CRIFO, C., BOZZI, A., MONDOVI, B., ROSSI-FANELLI, A.: The biochemical mechanism of selective heat sensitivity of cancer cells. IV. Inhibition of RNA synthesis. Europ. J. Cancer 9, 103 (1973)
163. SZYBALSKI, W.: Effect of elevated temperatures of DNA and on some polynucleotides: denaturation renaturation and cleavage of glycosidic and phosphate ester bonds. In: Thermobiology. Rose, A.N. (ed.) London: Academic Press, 1967, p. 73
164. SZYBALSKI, W., OPARA-KUBINSKA, Z.: Cellular Radiation Biology. Baltimore: Williams and Wilkins Co., 1965
165. TECCE, G., TOSCHI, G.: A study of ribosomes from a thermophilic microorganism. Suppl. Nuovo Cimento 18, 207 (1966)
166. THOMPSON, L.H., BAKER, R.M.: Isolation of mutants of cultured mammalian cells. In: Methods in Cell Physiology. Prescott, D.M. (ed.) New York: Academic Press, Vol. VI, 1973, p. 209
167. TONIOLO, D., MEISS, H.K., BASILICO, C.: A temperature-sensitive mutation affecting 28 S ribosomal RNA production in mammalian cells. Proc. nat. Acad. Sci. (Wash.) 70, 1273 (1973)
168. TURANO, C., FERRARO, A., STROM, R., CAVALIERE, R., ROSSI-FANELLI, A.: The biochemical mechanism of selective heat sensitivity of cancer cells. III. Studies on lysosomes. Europ. J. Cancer 6, 67 (1970)
169. UDAKA, S., HORIUCHI, T.: Mitants of Escherichia coli having temperature sensitive regulatory mechanism in the formation of arginine biosynthetic enzymes. Biochim. biophys. Res. Commun. 19, 156 (1965)
170. VERMEL, E.M., KUZNETSOVA, L.B.: Hyperthermia in the treatment of malignant diseases (in Russian). Probl. Oncol. USSR 16, 96 (1970)
171. VOLLMAR, H.: Über den Einfluß der Temperatur auf normales Gewebe und auf Tumorgewebe. Z. Krebsforsch. 51, 71 (1941)
172. WAROCQUIER, R., SCHERRER, R.: RNA-metabolism in mammalian cells at elevated temperature. Europ. J. Biochem. 10, 362 (1969)
173. WEINBERG, R.A.: Nuclear RNA metabolism. Ann. Rev. Biochem. 42, 329 (1973)
174. WESTERMARK, N.: The effect of heat upon rat tumors. Scand. Arch. Physiol. 52, 257 (1927)
175. WESTRA, A., DEWEY, W.C.: Variation in sensitivity to heat shock during the cell cycle of Chinese hamster cells in vitro. Int. J. Radiat. Biol. 19, 467 (1971)
176. WHITE, H.R.: The heat resistance of Streptococcus faecalis. J. gen. Microbiol. 8, 27 (1953)
177. WILSON, G., ROSE, S.P., FOX, C.F.: The effect of membrane lipid unsaturation on glycoside transport. Biochim. biophys. Res. Commun. 38, 617 (1970)
178. WOOD, T.H.: Lethal effects of high and low temperatures on unicellular organisms. Advanc. biol. med. Phys. 4, 119 (1956)
179. WOODCOCK, E., GRIGG, C.W.: Repair of thermally induced DNA breakage in Escherichia coli. Nature (New Biol.) 237, 76 (1972)

180. YERUSHALMI, A., HAR-KEDAR, I.: Enhancement of radiation effects by heating of the tumor. Israel J. med. Sci. 10, 772 (1974)
181. YOSHIZAKI, F., OSHIMA, T., IMAHORI, K.: Studies on phosphoglucomutase from an extreme thermophile Flavobacterium thetmophilum HB8. I. Thermostability and other enzymatic properties. J. Biochem. (Tokyo) 69, 1083 (1971)
182. ZAMENHOF, S.: Effects of heating dry bacteria and spores on their phenotype and genotype. Proc. nat. Acad. Sci. (Wash.) 46, 101 (1960)
183. ZAMENHOF, S., GREEN, S.: Heat as agent producing high frequency of mutations and unstable genes in E. coli. Nature (Lond.) 182, 611 (1958)
184. ZEIKUS, J.G., BROCK, T.D: Protein synthesis at high temperatures: aminoacylation of t-RNA. Biochim. biophys. Acta 228, 736 (1971)
185. ZEIKUS, J.G., TAYLOR, M.W., BROCK, T.D.: Thermal stability of ribosomes and RNA from Thermus aquaticus. Biochim. biophys. Acta 204, 512 (1970)

Chapter 3
Actions of Hyperthermia on Tumor Cells Cultured in vitro

B. C. GIOVANELLA

The first recorded observation that cultured neoplastic cells have higher thermosensitivity than normal cells was made in 1912.

LAMBERT (17), utilizing one of the first tissue culture systems, found that mouse and rat sarcoma cells are more thermosensitive than normal mesenchymal cells. For each temperature tested in the range 42°-47°C, it was possible to find a time of exposure which would suffice to kill sarcoma cells and yet allow mesenchymal cells to survive. Survival was measured by the capacity of cells to migrate from fragments of tissue maintained in a plasma clot. Curiously enough, this crucial observation elicited no interest and no follow-up.

In 1928 FRIEDGOOD (4), exposed hanging drop cultures of Walker rat tumor No. 1 to various temperatures for 10-30 min, incubated them at 38°C for 48 h, and then assessed their viability. Viability was once again determined by the presence or absence of cells migrating from the explant. The author assumed that only spindle-shaped cells were tumor cells, whereas the round cells observed were considered normal (mononuclear cells, probably macrophages). He did not find any difference in thermosensitivity between the two cell types. The obvious criticism of this experiment is that to differentiate between neoplastic and non-neoplastic cells in the same culture on purely morphologic grounds is impossible. In 1941, VOLLMAR (30), compared the survival at supranormal temperatures of normal (spleen cells, presumably fibroblasts) and neoplastic (Ehrlich ascites carcinoma and Jensen sarcoma) rodent cells. Tumor cells were killed at temperatures between 40° and 42°C, whereas normal cells resisted exposure upt to 43°C. Her experiments were carefully executed on a large group of cultures, but unfortunately quantitation of the results was very primitive, being based on subjective assessments of the rate of growth.

In 1966, BENDER and SCHRAMM (2), published the results obtained after heating for 30 min at various temperatures cultures of 46 different animal and human tumors. As controls, cultures of whole embryos and of various organs from six species were also tested. Survival was measured by further proliferation at 37°C. It was found that tumor cells were 1-2°C more thermosensitive than normal cells. Unfortunately, the authors did not count the cells before and after treatment, and thus it is difficult to estimate the quantitative effect of heating.

AUERSPERG (1) heated cultures of human fibroblasts and epithelial cells from carcinoma lines C4 and C27 at temperatures varying from 44 to 48°C for 30 min. She found that the neoplastic epithelial cells survived better than fibroblasts. It must be noted, though, that all the determinations of viability were by the indirect methods of vital staining and protein content which were afterwards demonstrated to be unreliable for the determination of cell killing caused by exposure to supranormal temperatures (5, 11, 19).

OSSOVSKI and SACHS (21) found that hamster cells transformed by polycyclic hydrocarbons had the same thermosensitivity as the cell of origin and that cells transformed by SV40 virus were more thermoresistant than both. None of these cells was tested for their tumor-producing ability.

KACHANI and SABIN (14) repeated the experiments of OSSOVSKI and SACHS and found that hamster cells transformed in vitro by SV40 or polyoma virus were as thermosensitive as normal cells, but if they were injected into hamsters and produced tumors, cultures derived from these tumors were more thermosensitive than normal hamster embryo cells. This observation can be easily explained if we postulate that a large number of the cells morphologically transformed in vitro were non-tumor-producing and that the in vivo passage was selective for the tumor producing cells, which were also highly thermosensitive.

In 1969 CHEN and HEIDELBERGER (3) found that mouse prostate cells cultured in vitro became more thermosensitive after undergoing malignant transformation after treatment with polycyclic aromatic hydrocarbons (Fig. 1).

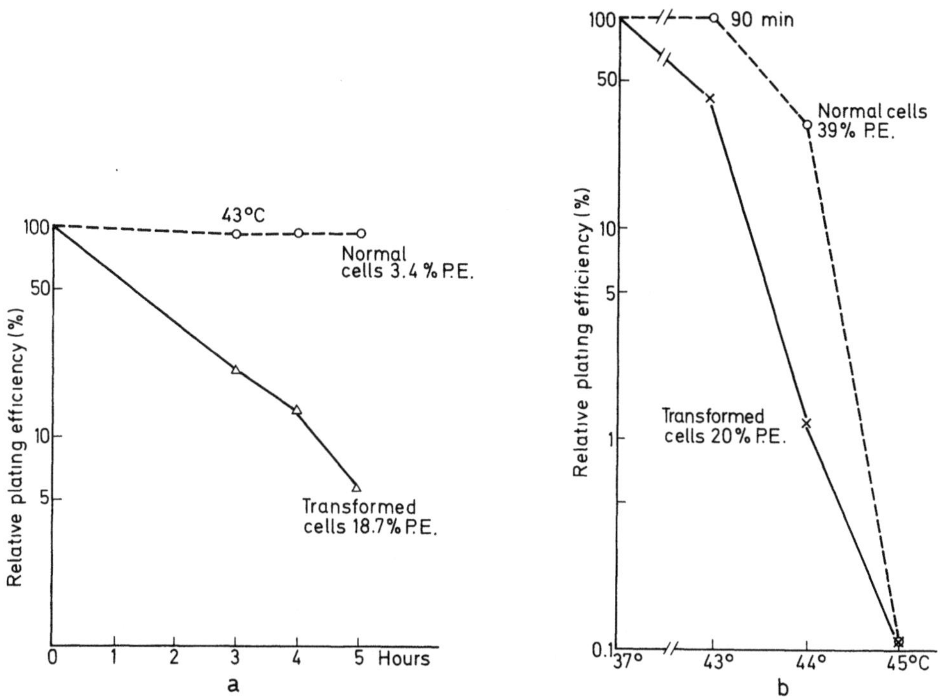

Fig. 1. A) Effect of heat treatment at 43°C on relative plating efficiencies of control and transformed cells. B) Effect of heating for 90 min at various temperatures on relative plating efficiencies of control and transformed cells. (From CHEN, T.T., HEIDELBERGER, C.: Int. J. Cancer 4, 166, 1969)

In this paper, the definition of "neoplastic cell" is supported by biological experiments. Ten transformed cells produced malignant tumors when injected in syngeneic mice whereas 10^3 "normal" cells did not.

LEVINE and ROBBINS (18) also compared the thermosensitivity of normal and neoplastic cells in culture, and found that the neoplastic ones were much more affected by supranormal temperatures. Unfortunately, their findings were based on an experimental set-up which could give only ambiguous results. Before heating, they allowed cells to reach confluence. At this point, normal, contact-inhibited cells stop multiplying whereas neoplastic cells continue to divide. Accordingly, the comparison was between neoplastic cells cycling and normal cells stopped in the Go phase. It has been abundantly demonstrated that the same cell changes its thermosensitivity according to the phase of the cell cyle during which the heating is performed (6, 22, 32). Both normal and neoplastic cells possess this property and the difference in thermosensitivity exhibited by the same cell heated during Go or cycling is very large. In order to have meaningful comparisons between a normal and a tumor cell, both must be heated when in logarithmic growth when their thermosensitivity is maximal. Theoretically, comparisons made with both cells in Go are also valid, but in practice it is difficult to make sure that all the tumour cells are in this stage.

In 1973, GIOVANELLA et al. (5) compared the thermosensitivity of normal and neoplastic mouse cells. The definition of normal and neoplastic cells was supported by inoculation into syngeneic hosts. The "neoplastic" cells (cultured from chemically induced fibrosarcomas) produced tumors when injected at doses of 1×10^6 cells in adults and 1×10^5 in newborns. The "normal" cells (fibroblasts cultured from mouse embryos of the same inbred strain C57 B1/6) did not. All the cultures were heated during their logarithmic growth phase and the doubling times of some of the "normal" cells were much shorter than those of the neoplastic cells. Without exception, all the "normal" cultures proved to be more thermoresistant than the neoplastic ones (Fig. 2). One of the "normal" cultures developed into an established cell line, MEF 1, lacking neoplastic properties, fast growing, but strictly contact-inhibited. From this line, a neoplastic subline, MEF 1T, was derived. These two lines have approximately the same doubling time (20-22 h), but whereas MEF 1T can produce tumors when 1×10^3 cells are injected in adult C57 B1/6 mice, 1×10^7 cells of MEF 1 did not produce any. Two hours of exposure to 43.5°C will kill 92-98% of MEF 1T cells and only 10-12% of MEF 1. These results, together with the work of CHEN and HEIDELBERGER (3), prove that mouse cells, as soon as they acquire malignant properties, become more thermosensitive than the cells of origin.

In 1975, KASE and KAHN (15) found that human nonneoplastic fibroblasts can be transformed by SV40 virus into cells with neoplastic potential as demonstrated by their ability to produce tumors when injected into nude thymus-deficient mice. It has been affectively demonstrated by GIOVANELLA et al. (7, 9, 10) that only human cultured cells from malignant neoplasms will grow as tumors when inoculated in "nude" thymus-deficient mice. Human or murine cells derived from normal tissues of adults or embryos will not, even if growing vigorously in vitro. Inoculation of tissue-cultured cells in nude thymus-deficient mice is, consequently, a sensitive test for their malignant potential. The human cells transformed by SV 40 virus increased their thermosensitivity considerably at the same time that they developed neoplastic capability.

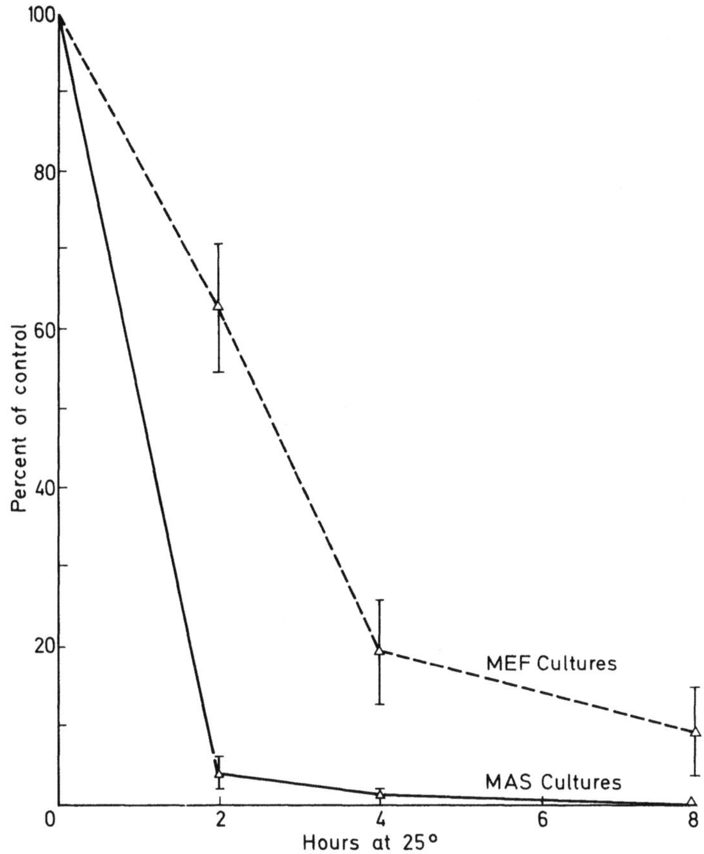

Fig. 2. Percentage of surviving mouse adult sarcoma (MAS) cells and normal mouse embryo fibroblasts (MEF) as function of time of exposure to 42.5°C. (Mean ± S.E. of pooled experiments are given.) At 2, 4, and 8 h, difference in heat sensitivity between MEF cells and MAS cells is statistically significant (P < 0.01). Seven MEF and 8 MAS lines were used. (From GIOVANELLA, B.C., et al.: Cancer Res. 33, 2568, 1973)

In 1975, STEHLIN et al. (28) also determined that human normal melanocytes derived from the uvea were more thermoresistant than human melanoma cells (Fig. 3). The same authors then (8) extended these findings to a wide variety of human nonneoplastic and neoplastic cells -- the definition "neoplastic" being given only to cells cultured from human neoplasms that are aneuploid and capable of producing malignant tumors when injected in "nude" thymus-deficient mice. The tumors were identified as human by karyotypic analysis. In this way, it was demonstrated that human colon carcinoma cells are more thermosensitive than normal epithelial cells derived from fetal intestine, that fibrosarcoma cells are more thermosensitive than normal fibroblasts, and so on. All the normal and neoplastic cells studied had similar doubling times; sometimes the normal cells divived faster than their neoplastic counterparts.

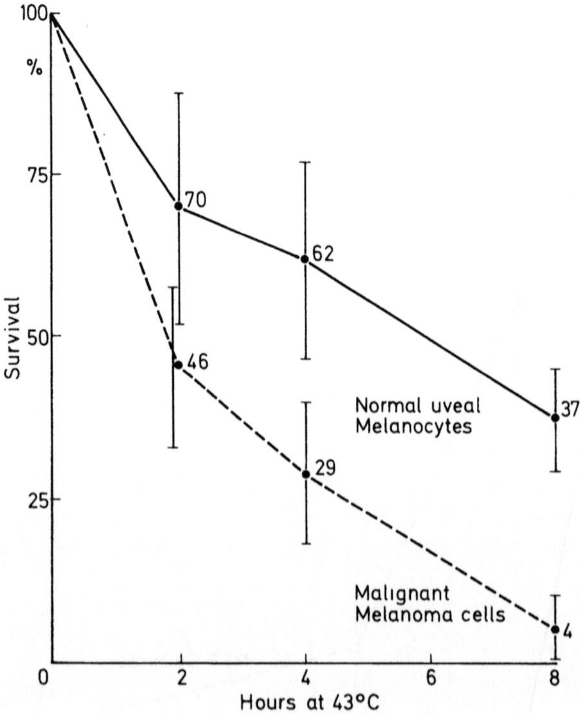

Fig. 3. Pooled data obtained using 11 lines derived from 6 human melanoma patients and 4 long-term cultures of fetal human uveal melanocytes. Results are statistically significant ($P < 0.01$) at 4 and 8 h. (From STEHLIN, J.S., et al.: Surg. Gynecol. Obstet. 140, 338, 1975)

All the work so far reported was done in order to answer the crucial question: Are nonneoplastic cells more thermoresistant than their neoplastic counterparts? The answer is yes, with the important qualification that the difference is not large and it is limited to a very small temperature range (41-43°C for human and rodent cells).

In recent years, some papers have appeared describing the thermosensitivity of normal cells in culture (12, 13, 24, 25, 27), and others describing this property in tumor cells (23, 26), with no comparisons being made between the two cell types.

Very little is known about what causes the difference in thermosensitivity of normal and tumor cells. This is not surprising because, in the first place, we do not know why cells (both normal and neoplastic) die after exposure for a certain time to supranormal temperatures, up to 44-45°C for mammalian cells. If the limiting factor is the same for both normal and neoplastic cells, with only a quantitative difference, or if in the tumor cell a new, more thermosensitive vital function is present, is a question that cannot be answered today.

We need to increase our knowledge of the physiology and pathology of supranormal temperatures at the cellular level. A very promising field of study for these purposes is the analysis of temperature-sensitive mutant cells (16, 20, 29, 31) compared to the lines of origin.

REFERENCES

1. AUERSPERG, N.: Differential heat sensitivity of cells in tissue culture. Nature (Lond.) 209, 415-416 (1966)
2. BENDER, E., SCHRAMM, T.: Untersuchungen zur Thermosensibilität von Tumor- und Normalzellen in vitro. Acta biol. med. ger. 17, 527 (1966)
3. CHEN, T.T., HEIDELBERGER, C.: Quantitative studies on the malignant transformation of mouse prostate cells by carcinogenic hydrocarbons in vitro. Int. J. Cancer 4, 166 (1969)
4. FRIEDGOOD, H.B.: On the thermal death point of sarcoma and normal mononuclear cells (Walker rat tumor No. 1). Arch. exp. Zellforsch. 7, 243 (1928)
5. GIOVANELLA, B.C., MORGAN, A.C., STEHLIN, J.S., WILLIAMS, L.J.: Selective lethal effect of supranormal temperatures on mouse sarcoma cells. Cancer Res. 33, 2568 (1973).
6. GIOVANELLA, B.C., MOSTI, R., HEIDELBERGER, C.: Further studies of the lethal effects of heat on tumor cells. Proc. Amer. Ass. Cancer Res. 10, 29 (1969)
7. GIOVANELLA, B.C., STEHLIN, J.S.: Assessment of the malignant potential of cultured cells by injection in "nude" mice. In: Proc First Internat. Workshop on Nude Mice. Rygaard, I., Povlsen, C.O. (eds.) Stuttgart: Fischer Verlag, 1974, pp. 279-284
8. GIOVANELLA, B.C., STEHLIN, J.S., MORGAN, A.C.: Selective lethal effect of supranormal temperatures on human neoplastic cells. Cancer Res. 36, 3944 (1976)
9. GIOVANELLA, B.C., STEHLIN, J.S., WILLIAMS, L.J., Jr.: Development of invasive tumors in the "nude" mouse after injection of cultured human melanoma cells. J. nat. Cancer Inst. 48, 1531 (1972)
10. GIOVANELLA, B.C., STEHLIN, J.S., WILLIAMS, L.J., Jr.: Heterotransplantation of human malignant tumors in "nude" thymusless mice. II. Malignant tumors induced by injection of cell cultures derived from human solid tumors. J. nat. Cancer Inst. 52, 921 (1974)
11. HARRIS, M.: Criteria of viability in heat-treated cells. Exp. Cell Res. 44, 658 (1966)
12. HARRIS, M.: Temperature-resistant variants in clonal populations of pig kidney cells. Exp. Cell Res. 46, 301 (1967)
13. HARRIS, M.: Growth and survival of mammalian cells under continous thermal stress. Exp. Cell Res. 56, 382 (1969)
14. KACHANI, Z.F.C., SABIN, A.B.: Reproductive capacity and viability at higher temperatures of various transformed hamster cell lines. J. nat. Cancer Inst. 43, 469 (1969)
15. KASE, K., HAHN, G.M.: Differential heat response of normal and transformed human cells in tissue culture. Nature (Lond.) 255, 228 (1975)
16. KIMURA, G.: Temperature sensitive growth of cells transformed by mutant of polyoma virus. Nature (Lond.) 253, 639 (1975)
17. LAMBERT, R.A.: Demonstration of the greater susceptibility to heat of sarcoma cells. J. Amer. med. Ass. 59, 2147 (1912)
18. LEVINE, E., ROBBINS, E.R.: Differential temperature sensitivity of normal and cancer cells in culture. J. cell. Physiol. 76, 373 (1969)
19. MUCKLE, D.S., DICKSON, J.A.: The selective inhibitory effect of hyperthermia on the metabolism and growth of malignant cells. Brit. J. Cancer 25, 771 (1971)
20. NAHA, P.M.: Temperature sensitive conditional mutants of monkey kidney cells. Nature (Lond.) 223, 1380 (1969)
21. OSSOVSKI, L., SACHS, L.: Temperature sensitivity of polyoma virus: Induction of cellular DNA synthesis and multiplication of transformed cells at high temperature. Proc. nat. Acad. Sci. (Wash.) 58, 1938 (1967)

22. PALZER, R.J., HEIDELBERGER, C.: Influence of drugs and synchrony on the hyperthermic killing of HeLa cells. Cancer Res. 33, 422 (1973)
23. PALZER, R. J., HEIDELBERGER, C.: Studies on the quantitive biology of hyperthermic killing of HeLa cells. Cancer Res. 33, 415 (1973)
24. ROSCOE, D.H., READ, M., ROBINSON, H.: Isolation of temperature sensitive mammalian cells by selective detachment. J. cell. Physiol. 82, 325 (1973)
25. ROSCOE, D.H., ROBINSON, H., CARBONELL, A.W.: DNA synthesis and mitosis in a temperature sensitive Chinese hamster cell line. J. cell. Physiol 82, 333 (1973)
26. SELAWRY, O.S., GOLDSTEIN, M.N., McCORMICK, T.: Hyperthermia in tissue-cultured cells of malignant origin. Cancer Res. 17, 785- (1957)
27. SMITH, B.J., WIGGLESWORTH, N.M.: A temperature-sensitive function in a Chinese hamster line affecting DNA synthesis. J. cell Physiol. 82, 339 (1973)
28. STEHLIN, J.S., GIOVANELLA, B.C., de IPOLYI, P.D., MUENZ, L.R., ANDERSON, R.F.: Results of hyperthermic perfusion of melanoma of the extremities. Surg. Gynec. Obstet. 140, 338 (1975)
29. THOMPSON, L.H., MANWOVITZ, R., BAKER, R.M., TILL, J.E., SIMINOVITCH, L., WHITMORE, G.F.: Isolation of temperature sensitive mutants of L cells. Proc. nat. Acad. Sci (Wash.) 66, 377 (1970)
30. VOLLMAR, H.: Über den Einfluß der Temperatur auf normales Gewebe und auf Tumorgewebe. Z. Krebsforsch. 51, 71 (1941)
31. WEISS, H.K., BASILICO, C.: Temperature sensitive mutants of BHK 21 cells. Nature (New Biol.) 239, 66 (1972)
32. WESTRA, A., DEWEY, W.C.: Variation in sensitivity to heat shock during the cell cycle of Chinese hamster cells in vitro. Int. J. Radiat. Biol. 19, 467 (1971)

Chapter 4
The Effects of Hyperthermia in Animal Tumour Systems

J. A. DICKSON

1. INTRODUCTION

Heat, in the form of red-hot iron (ferrum candens) and chemical caustics, was the most commonly employed modality for the treatment of cancer in ancient times. The cauterization of malignant tumours by ferrum candens was practised as long ago as 2000 B.C., and was recommended by Hippocrates for small, non-ulcerating tumours (187). The first documented indication that the destructive effects of elevated temperature might be selective for cancer cells is usually ascribed to BUSCH, who in 1866 reported the disappearance of a histologically verified sarcoma of the face after two attacks of erysipelas (17). BRUNS (14) and COLEY (26) then noted that patients suffering from neoplastic disease often had a regression of both primary and secondary tumours following an attack of erysipelas or other severe pyrexial infection. Attempts were soon made to take advantage of these findings for the treatment of cancer using locally or systemically applied heat in various forms, and remissions were described in patients with a variety of tumours, including cancer of the breast, face, cervical region, penis, and also in malignant melanoma, periosteal sarcoma, lymphosarcoma, ovarian and uterine carcinoma (see 19, 22, 175 for refs.). These clinical experiences generated widespread interest in hyperthermia in the early years of this century, and the destructive effects of heat on human tumours were confirmed both *in vivo* and *in vitro* for transplantable animal tumours; the relative resistance of normal tissues to damage by elevated temperature was also corroborated (19, 175).

By the onset of the Second World War, therefore, there had accumulated a considerable literature that was suggestive of the possible potential of hyperthermia as an antitumour agent. Hints that this potential might reside in a selective susceptibility of cancer cell biochemistry to heat, thus offering a metabolic Achilles heel for tumour therapy, had also been mooted. As long ago as 1927, WESTERMARK (183) showed that, for the Flexner-Jobling carcinoma and the Jensen sarcoma in the rat, the time required at different temperatures (44-48°C) to obtain tumour regression *in vivo* corresponded well with the time necessary *in vitro* to reduce respiration and glycolysis (aerobic and anaerobic) of tumour slices to zero. Compared to the malignant tissue, at these temperatures a longer time was required to inhibit respiration in normal rat liver and spleen.

In spite of this large body of evidence attesting that heat can adversely influence tumour growth, and notwithstanding more recent confirmatory results with heat in patients with cancer (19, 114) and with a wide range of transplantable (30, 41, 122, 127) and also spontaneous tumours in animals (29, 144), hyperthermia has been slow to gain recognition as a possible modality for the armamentarium of the cancer therapist. There are several reasons for this. Many of the earlier clinical reports were evaluated on the personal experience of the investigator rather than against objective criteria. Much of the work on tumour destruction in animals referred to applied temperatures that were nearer 50°C than 40°C. For the treatment of cancer in animals or in humans by total body heating, temperatures in excess of 42°C are not feasible because of host tolerance considerations (35, 38); for local hyperthermia, reliable techniques are not yet available for the selective generation of intra-tumour temperatures in the region of 50°C. A myriad of techniques has been employed over the years to heat tumours, including local heating with fomentations (66, 134), water bath immersion (28, 62, 171), shortwave diathermy (90, 127), microwave diathermy (144, 183), ultrasound (152), hyperthermic regional perfusion (19, 165, 171), specially designed hotboxes (178) and more recently, induction of total body hyperthermia by inhalation of warmed anaesthetic gases with insulation against heat loss by encasing the body in wax (78). The most widely applied technique clinically was that of COLEY, who induced pyrexia in patients with cancer by the injection of bacterial filtrates, which became known as COLEY's toxins. In spite of the considerable success achieved by COLEY in curing various types of advanced malignant disease over many years (125), the bacterial preparations proved impossible to standardize biologically, and consequently their effects were capricious and inconsistent. The method fell into general abeyance following COLEY's death in 1936, although sporadic reports of its beneficial use still appear (114).

The lack of adequate control data in the older clinical material, the numerous heating techniques employed, and the almost complete failure prior to 1940 to record temperature in the heated tumours (the sophisticated work of WESTERMARK (183) is the notable exception) has rendered data difficult to evaluate, repeat, or to formulate into a standardized approach for heating tumours. The potential of hyperthermia for the treatment of cancer has therefore not been realized to date. In spite of the considerable literature describing the deleterious effects of elevated temperature on cancer cells, few guidelines have emerged for therapy. There is little precise information available concerning the optimal conditions of heating (exact temperature to be used, duration of application), the susceptibility of various tumours, and the relative effectiveness of heat applied to the tumour (local hyperthermia) compared with total body heating.

In 1968 we began to investigate some of these problems in animal model systems, and it is the purpose of the present chapter to detail the results to date. The work has concerned the response of the tumour-bearing host, as well as the tumour, to hyperthermia. "Around every tumour there is a patient" (11) is not only a maxim but a fact emphasized strongly by GREENSTEIN in the 1950s (71) to stress the importance of the host-tumour relationship in cancer. Clinical experience suggests that each individual has his own resistance against his own cancer, and that duration of survival may be influenced by this resistance or immunity (169), thus reiterating the significance of the host component in the host-tumour relationship. It is the opinion of the present author that much of the work on hyperthermia in relation to tumours has ignored the host, and consequently the value of the results obtained in relation to human neoplasia is open to question. There is increasing evidence that the pathogenesis of a given neoplasm is governed by a balance operating between the growing tumour and the reaction of the host (see below). Studies which refer only to tumour response in therapy are of limited value, therefore. The effects of hyperthermia will be discussed against the background of host-tumour interaction and in relation to current concepts of tumour immunology, as far as this seems relevant at the present time.

2. ANIMAL TUMOUR MODELS

As in other aspects of research employing animal models, the value of results in the present work is governed by the limitations of the experimental system. The usefulness and shortcomings of transplantable tumours have been reviewed by several authors (see 102 for refs.) The speed and relative ease of obtaining reasonably homogenous samples of malignant tissue, growing at a predictable rate and available in large quantities, has led to the preferential use of transplantable tumours in cancer research. It is accepted that with a transplantable tumour the host animal is suddenly confronted with an overwhelming number of cancer cells that produce rapid and dramatic changes leading to the appearance of a tumour. A spontaneous tumour, on the other hand, arises as the result of more subtle and slower changes (102). It has been pointed out recently that an important similarity between transplanted tumours and human tumours is neoantigen formation and antigen deletion that occurs as selection of tumour growth against immune resistance takes place (95). Metastasis is not nearly so conspicuous a feature of malignant neoplasia in experimental animals as it is in man (51, 168); however, in general, malignant tumours of other mammals metastasize in ways similar to those of man (185). If we are to establish guiding principles for the treatment of human cancer, it is imperative that our animal model system should be as good a replica as possible of the disease in humans. This implies using a tumour that metastasizes. Most cancer patients do not die of the local growth; they die of distant metastases that may or may not correlate with the size of the local growth. Metastasis is the sine qua non of cancer, and the development of metastasis is governed by the tumour-host relationship. Many of the past publications in the field have concentrated on the effects of heat on primary tumours that did not metastasize. Such models are of value for investigating methods of applying heat to tumours - that is, methods of tumour cure. This is not equated with host cure and hence the value of the results in relation to human cancer is considerably diminished.

Transplantation techniques have been discussed in detail by LIEBELT and LIEBELT (102). Much valuable information regarding the study of metastasis in animal systems in available in the review by FISHER and FISHER (48). From the point of view of the present work, the following points are of importance. An intramuscular inoculation of the cancer cells studied gave rise to a tumour that consistently metastasized; the primary mass was convenient for measurement and for heating. Subcutaneously transplanted tumours do not commonly metastasize (102). It is not widely appreciated that after subcutaneous inoculation, the majority of tumour cells do not remain at the injection site; as many as 70% of the cells may depart from the site within 24 h (49). This constitutes a major criticism of the system, since metastases may be difficult to distinguish from multiple primary foci of tumour cells. Primary tumour growth and the incidence of metastasis are roughly proportional to the number of cells inoculated, and may be governed by the strain, sex and age of the host animal used (48). Solid tumours undergo necrosis with progressive growth; the pattern and degree of necrosis varies with different neoplasms, but with transplantable animal tumours is usually fairly constant in time of appearance and pattern for a given cell inoculum. Detailed histological studies of tumour growth are therefore required if therapy is to have meaningful results. The response of solid tumours to therapy is governed by the kinetics of the tumour cell population, tumours being most susceptible to therapy (except surgery) during the exponential phase of the growth curve (160).

These considerations emphasize the prime importance of standardizing the experimental system from the point of view of host animals used, number of viable tumour cells inoculated, growth characteristics of the tumour and experimental design. The pattern of growth and metastasis for each individual tumour contemplated as a model system should be determined prior to establishing the experimental protocol. Although most well known transplantable tumours now in use have achieved a stable histology and growth pattern over the years, it must be remembered that such tumours are not static in this respect; alterations in growth rate and incidence of metastasis can occur (102). Such factors should be kept in mind and as far as possible accounted for in the interpretation of experimental results. A further point requiring consideration in the choice of tumour system is that some of the classic transplantable tumours (e.g. Jensen sarcoma and Flexner-Jobling carcinoma) can have spontaneous regression rates as high as 20-25% (168). It must be emphasized, however, that spontaneous regression, if it occurs, is a feature of the exact tumour type and line used as well as the particular animal host (and even the particular animal supplier). In addition, the biological behaviour of a tumour may vary in different laboratories. It is therefore the responsibility of the individual experimenter to define these basic characteristics of his animal tumour model if reproducible and meaningful results are to be obtained.

3. CURRENT APPROACHES

The present revival of interest in hyperthermia in relation to cancer is due in large measure to the reports of ROSSI-FANELLI and co-workers (19, 117, 119), that the respiration and nucleic acid and protein synthesis of Novikoff hepatoma and Ehrlich ascites carcinoma cells was selectively and irreversibly inhibited at 42°C in vitro, while the metabolism of normal and regenerating rat liver cells was unaffected by the elevated temperature. The Italian group proceeded to illustrate the potential for exploiting this difference in biochemistry between normal and cancer cells by treating cancers of the limbs by regional perfusion with heated blood (19). Tumour temperature was maintained in the region of 41.5-43.5°C for up to 8 h, and in 15 of 22 patients, there was massive necrosis and disappearance of the tumours. In the present work, an *in vitro*-*in vivo* approach has been used to investigate the extent to which *in vitro* results can be used to predict tumour response in the host to heat.

The tumours studied were the VX2 carcinoma in the rabbit, an adenocarcinoma of the breast in the rat, and the Yoshida sarcoma in the rat. These are solid transplantable tumours that kill the host by metastasis, and by inoculation of known amounts of tumour into the leg muscles the time pattern of metastasis can be standardized. The VX2 carcinoma is an anaplastic derivative of the Shope virus papilloma (168), and metastasizes to the regional and aortic lymph nodes and lungs. The breast adenocarcinoma was induced by administration of a single gastric dose of 7,12 dimethylbenzanthracene to a Sprague-Dawley rat (85); the tumour is maintained in monolayer culture, and the cells metastasize widely from inoculation into any of several primary sites (39). The Yoshida sarcoma, another chemically induced tumour, arose after feeding an albino rat O-aminoazotoluol, followed by painting the skin with potassium arsenite (168). From a primary inoculation site in the leg, the Yoshida tumour spreads extensively via direct, lymphatic and vascular routes to involve adjacent tissues and the thoracic and abdominal viscera (36).

For the VX2 work, male New Zealand white rabbits weighing 2.5 kg were used; the bioassay system for the breast cancer cells involved the use of 4-6-week-old Sprague-Dawley rats of both sexes; the Yoshida tumour was studied in 200-250 g outbred Wistar rats of both sexes. All animals were fed an ad libitum diet. Under these conditions, and with the site and inoculum described, the spontaneous regression rate for the three tumours has been zero in our hands.

A. In vivo Methodology

The tumours were inoculated into one hind limb of the host, either as a single-cell suspension or as 0.1 ml (100 mg) tumour homogenate. Cell suspensions were prepared by trypsinization, and only preparations with a viability in excess of 90%, as judged by dye exclusion, were used (122). Tumour homogenate was obtained by dicing the material (kept moist in isotonic saline) into small fragments (less than 1 mm^3) with a scalpel blade -- the term 'homogenate' is not connected with the biochemical connotation of the word.

Tumour volume was calculated from caliper measurements made in the three orthogonal (antero-posterior, lateral and vertical) diameters, allowance being made for the thickness of each animal's own tissues from measurements made on the limb before tumour inoculation. For rapid assessment, the formula $V = 0.52\ cde$, where c, d and e are the diameters (136), gives a convenient estimate of tumour volume. For more accurate measurement of volume, the diameters at each point on the growth curve should be tested in the various formulae for geometric shapes (82) and the resultant volume compared to the corresponding volumes obtained by water displacement. In this way, the best approximation for tumour mass (sphere, ellipsoid) is obtained for construction of a tumour growth curve. The methods of plotting and analysing tumour growth and the relevant factors in the presentation of such data have been reviewed by MENDELSOHN and by LAIRD (34, 98, 111). The response of tumours to chemotherapy (161) and to radiotherapy (53) is governed by the kinetics of the tumour cell population, and such factors as cell cycle time, growth fraction and cell loss alter with tumour growth (56, 112, 164). For reproducible and meaningfull results, therefore, antitumour therapy should be identified in terms of the growth curve of the tumour being examined.

a. Tumour Heating

Many of the difficulties in the field of hyperthermia in cancer therapy have concerned, and still do concern, the method of applying heat to tumours. Numerous techniques of heating have been tried ranging from microwaves to regional perfusion (see earlier). Each method has its own inherent difficulties.

Local Hyperthermia. With microwaves, the heating effect is governed by the wavelength used; different degrees of transmission and absorption of the waves by various tissues (fat, muscle, bone tumour), and reflection of the waves at tissue boundaries, render this form of heating difficult to control (89). For reliable temperature measurements the thermocouple must be balanced for the high frequency field, and even when superficial tumours are heated, a number of uncontrollable factors render uniform heating of the tumour difficult to attain (127). Workers who have used high frequency currents for heating tumours have developed their own methods specifically designed to heat a tumour mass of a given size in a specific anatomical site (127, 183). With these currents it is theoretically possible to selectively heat tissues or tumours at various depths and in various anatomical sites (89, 104), and sophisticated computerized formulations have been devised for temperature distribution patterns in different tissues and at tissue interfaces (81). However, much research is still required before a generally applicable microwave method for heating tumours is developed (81, 89, 104). The use of ultrasonics has been hampered by the need to maintain some form of physical continuity between the probe and the tissue during treatment. Regional perfusion is a specialized surgical technique most suited to limbs and not readily applicable to animals smaller than dogs. Complications such as shock, heamorrhage, oedema and severe damage to the perfused limb can occur (19, 171). Even when performed as a major surgical procedure in man with measures such as blood transfusion to minimize complications (19), the great disadvantage of the method is that repeat perfusion is difficult due to fibrosis and formation of adhesions in the operative area. The simplest method of heating tumours is by water bath immersion, and foot tumours, leg tumours and subcutaneous tumours on the abdomen can be readily treated in this way. The method is convenient for the treatment of foot tumours in rodents, since the normal structures of the foot in rats and mice represent a sensitive indicator of normal tissue damage. For ease of application, capacity of heating several tumours simultaneously and homogeneously, and versatility of temperature range, water bath immersion is a reliable method of controlled hyperthermia for animal tumours.

The heating bath used in this laboratory consists of a perspex tank (33 x 33 x 15 cm) containing 10 l of water heated by a Circotherm II Constant Temperature Unit with a 700-watt coil heater and circulating pump with an output of 12 l/min (Shandon Scientific Co. Ltd., London, England). At an ambient temperature of 25°C this unit maintains the bath temperature constant to ± 0.05°C, as measured by three temperature probes placed at various positions in the bath. During heating of tumours two temperature sensors, one in each half of the bath, are connected to the same Lightmeter as the sensors used to monitor temperature in the tumour and in the animal. With this heating unit and bath the water temperature can be altered rapidly and easily; such control is essential, since with small tumours (on the foot of a rat, for example) the tumour temperature is closely related to the bath temperature. For heating rabbit tumours, the tumour-bearing limb is immersed in the bath to completely submerge the tumour; two rabbits can be heated at the same time. With rats, the animals are placed on a perspex platform resting over the bath and the tumour-bearing limb or foot is inserted into the water through a 10 cm diameter padded opening; the foot is supported at a depth that allows complete submersion of the tumour.

Total Body Hyperthermia. The rationale of total body heating is that by elevating the general body temperature it should be possible to destroy heat-sensitive cancer cells throughout the body. This approach has immediate appeal since dissemination and metastasis are the lethal aspects of this disease that have proved unyielding to therapy to date. The bulk of the literature on the treatment of cancer by hyperthermia concerns the application of heat locally to tumours; less information is available on total body heating. LOVELOCK and SMITH (106) have devised diathermy techniques for the rapid heating of whole frozen animals (hamsters and rabbits) without injury to the animal, and BIGELOW et al. (10) found that diathermy was applicable to rewarming of hypothermic patients. At temperatures of 37°C and above, however, a danger of this technique is the differential sensitivity of normal tissues to overheating (106). Total body immersion in specially designed water tanks has been used by several continental workers to treat patients with cancer (4, 175); the published results are not impressive, however. The main disadvantage of the hot bath method is that the temperature difference between the heating fluid and the skin cannot be large because of the danger of burning. Recently, HENDERSON and PETTIGREW (78) have devised an ingenious method of elevating body temperature in humans, using the large alveolar surface area of the lungs as heat exchanger. The patient's temperature is elevated by inhalation of a prewarmed mixture of 50% helium in oxygen, and heat loss is then minimized by encasing the body in paraffin wax. By this means a controlled elevation of body temperature to 42°C can be achieved rapidly and reproducibly, and has been maintained for up to 10 h. For total body hyperthermia in animals, water bath immersion and the HENDERSON-PETTIGREW technique require sophisticated anaesthetic techniques; the commonly employed agents (e.g. barbiturates) do not alone give the degree of controlled anaesthesia required, and involuntary movements can easily lead to drowning of the animal in water bath immersion (38).

The most convenient method for elevation of total body temperature in animals is radiant heating. This can be applied from electric light bulbs in a specially insulated hotbox (38), or a standard laboratory incubator can be used, provided there is access for the temperature monitoring probes.

b. Temperature Measurement

Accurate appraisal of the value of hyperthermia for cancer therapy requires a baseline for reference purposes. The natural candidate for this baseline is intra-tumour temperature. Measurement of intra-tumour temperature gives more meaningful results than reference to the applied temperature. The temperature differential between a limb tumour and the immersion bath or perfusion fluid may be as much as 3-4°C, for example, depending upon several factors, including the type and size of animal, limb size and blood flow, tumour volume and bath dimensions. With limb tumours, there is an induction period before intra-tumour temperature reaches a plateau. For these reasons, we define hyperthermia in terms of intra-tumour temperature.

Tumour heating may not occur homogeneously, and the more deeply situated cells may be slower to achieve a given temperature than superficial parts of the neoplasm. If the tumour abuts on bone, we have found that in this part of the neoplasm the temperature may run at a level lower than in the main tumour mass. By employing several sensors simultaneously inserted into the tumour, temperature differences and gradient patterns are determined in the tumour during induction and maintenance of the required elevated temperature. It can then be decided whether one electrode gives a reliable index of tumour temperature, or whether more than one electrode is required. In the latter case, the record of the lower-reading electrode is taken as the measure of tumour temperature.

The few previous workers who have measured intra-tumour temperature in animal studies have designed their own ironconstantan (183), or copper-constantan (90) thermocouples incorporated in a metal needle, or the thermocouple has been surrounded by a fine air-filled glass tube (126, 127). These probes were tedious in operation since they had to be calibrated and balanced during operation against a constant temperature junction (90, 183), nor did the galvanometers lend themselves easily to multiple simultaneous temperature recordings. In our earlier work we used copper-constantan thermocouple probes linked to a Cambridge potentiometer (type 44228), but we now favour the much more convenient direct reading electric thermometers. The instrument used in a 12-channel model (Model 3 GID, Light Laboratories, Brighton, England) with a scale range of 36°-46°C; it has a fast response time of 4 s, records temperature with an accuracy of ± 0.05°C, and is unaffected by ambient temperature changes. We routinely employ two of these thermometers that have been simultaneously standardized and calibrated for interchangeable use with a variety of probes.

For intra-tumour or intra-abdominal temperature measurement the thermistor probes used are 5-cm-long needle-type IH 0.8 mm in diameter, recording temperature change only at the tip; polythene-covered type IMR for rectal measurement; and immersion type IRM for insertion into the bath. With the needles, the depth of insertion depends upon tumour size and number and positions of probes per tumour; generally, the aim is to record temperature with the needle tip in the central region of the tumour. Central body or "core" temperature is monitored by an intra-abdominal needle inserted below the liver in a right paramedian position and by a polythene or rubber rectal probe. Again, depths of insertion depend upon animal type and size. For a 200-250 g rat, for example, the abdominal probe is inserted 2.5 cm through the skin, and the rectal probe 3 cm beyond the anus. In the rat, a rectal probe (well coated first with lubricating jelly) can be inserted 10 cm beyond the anus. At this depth, the thermistor tip of the probe is usually situated in the hepatic flexure of the colon in proximity to the undersurface of the liver; the probe can be palpated in the animal's abdomen and easily guided into the region of the liver. In this position it has been found that the probe records a temperature similar to the intra-abdominal needle, i.e. "core" temperature. A deeply inserted rectal probe is therefore a convenient means of monitoring body temperature. Probes can often be splinted in position to prevent dislodgement during heating -- in the rat the rectal probe can be taped to the base of the animal's tail, and with foot tumours the needle electrode can be immobilized by a non-restricting tape bandage round the lower leg.

For total body heating, the environmental temperature in the hotbox or incubator is continously monitored by one or more type 1AG air-gas probes connected to the Light electric thermometer.

c. Animal Care

For hyperthermia experiments, the animals are anaesthetized with Nembutal veterinary (60 mg pentobarbitone sodium per ml, Abbot Laboratories, Queensborough, Kent, England). Rabbits are given 0.6 ml/kg body weight intravenously. Rats receive 0.1 ml of a 1:5 dilution of the commercial solution per 50 g body weight i.p. During the heating procedure it is often necessary to administer further Nembutal as the drug is metabolized more rapidly at increased temperature and the depth of anaesthesia becomes less.

Several important points concerning the host animal require consideration. Following injection of the anaesthetic, the temperature of the animals can fall rapidly while the electrodes are being placed in position. The temperature of a rat may fall to the region of 30°C, for example, and this can cause consideralbe delay in achieving an elevated temperature. At the end of the experiment, the animal's temperature may again fall precipitously on removal from the bath or hotbox. Such swings in temperature can be avoided by maintaining a fairly high room temperature (in the region of 25°C) and by keeping the animals warm while they are anaesthetized but not being heated. Before hyperthermia, it is convenient to place the animals on the perspex platform over the heating bath or under an infra-red heater while the electodes are positioned. After heating, the animals can be wrapped in a blanket and placed under an infra-red heater for 10-15 min; this helps to control the return of body temperature to normal without rapid overswing to subnormal levels.

During and immediately after heat therapy, be it by water bath immersion or by radiant heat, fluid loss must be replaced in the animals. This is achieved by i.p. injection of 4% dextrose-0.18% NaCl to a total of approximately 20% of the animal's blood volume (total blood volume of the animal is calculated as equal to 4.5% body weight (147). For local hyperthermia lasting 60-90 min, total fluid replacement volumes of 2.5 ml in a 250 g rat and 25 ml in a 2.5 kg rabbit are adequate; for systemic hyperthermia, these volumes are increased by 50%.

B. In vitro Methodology

a. Respiration and Glycolysis

Almost 50 years ago, WESTERMARK (183) reported that both the respiration and glycolysis in slices of Flexner-Jobling carcinoma and Jensen sarcoma ceased after heating at 45°C for 90 min. At higher temperatures, a proportionally shorter exposure time was required to inhibit these parameters. WESTERMARK found that the lengths of time required to inhibit respiration and glycolysis *in vitro* were well in accord with the corresponding times necessary for obtaining recession of the tumours by heat treatment at these temperatures. These findings suggest the possibility of using *in vitro* assay systems to predict the *in vivo* response of tumours to hyperthermia. More recently, CAVALIERE et al. (19) and MONDOVI et al. (119) have confirmed the inhibitory effect of elevated temperature (42°C) on cancer cell respiration using Ehrlich ascites carcinoma cells and hepatoma cells of the Novikoff and Morris 5123 lines. In addition, these workers reported that the incorporation of labelled precursors into DNA, RNA and protein of the cancer cells was inhibited at 42°C (117). From the results with these cells, MONDOVI et al. postulated that the damage by heat to respiration in tumour cells should be regarded as a general phenomenon (119). Recent work with other types of cancer cells has corroborated the inhibitory effect of heat on respiration (39, 41, 122) and on nucleic acid and protein synthesis (39, 41, 130), and has in general supported this postulate of MONDOVI et al. The notable and significant exceptions to the hypothesis would appear to be human tumours, in which glycolysis can be inhibited while respiration remains unimpaired (40). Using several types of tumour cell and associated *in vitro-in vivo* systems, we have found that irreversible inhibition of tumour respiration at 42°C and the simultaneous curtailment of nucleic acid and protein synthesis are correlated with inability of the cells for further replication *in vitro* (39) and failure to produce tumours on transplantation (39, 41, 122). We have therefore adopted the measurement of respiration and glycolysis as a rapid means of assessing the *in vivo* response of a tumour to hyperthermia.

From studies on the thermal death times of Jensen sarcoma and Walker 256 carcinoma slices, JARES and WARREN (87) and JOHNSON (90) emphasized the fundamental importance of technique for *in vitro* studies on hyperthermia. These workers found that small differences in the conditions of experiments (use of balanced saline solutions rather than physiological NaCl for tissue incubation, adequate shaking of suspended slices for oxygenation) gave rise to results that were due to factors other than the heating effect. For this reason, JOHNSON (90) criticized the earlier results of ROHDENBURG and PRIME (144) on thermal death times of sarcoma 180 as being too short owing to the use of unphysiological conditions of measurement. MONDOVI et al. (119) recently illustrated the importance of the heating conditions for *in vitro* experiments by demonstrating a difference in manometric measurements made at elevated temperature, and results obtained at 38°C after preincubation at the increased temperature. When cells of the minimal deviation hepatoma 5123 were incubated at 38°C and 43°C, there was no difference in oxygen consumption of the suspensions; if the cells were incubated at 43°C and then Warburg manometry performed at 38°C, oxygen uptake was reduced.

We are faced, therefore, with the choice of subjecting the tumour tissue to hyperthermia during a pre-incubation period followed by metabolic measurements at normal body temperature, or performing Warburg manometry at elevated temperature (direct incubation) and then again at body temperature. The crucial factor is whether or not any inhibition of respiration observed at, say 42°C, is irreversible at normal body temperature. This can only be ascertained by returning the heated cells to 38°C. With most heat sensitive cancer cells from animal tumours that have been studied in detail, there has been a total inhibition of O_2 uptake after 1-4 h at 42°C (19, 39, 41, 119, 122, 183); this time interval becomes progressively less at temperatures above 42°C (122, 183). Such cells, in which respiration has ceased at 42°C, may continue to respire at a low level for 2 h or more when returned to 38°C; the cells have lost their replicative potential, however, and do not form tumours on transplantation into host animals (39, 41, 183). If the heating time is inadequate to cause total inhibition of O_2 consumption, it can readily be shown that the inhibition is reversible at 38°C and the cells produce tumours on inoculation into animals (122). Recently, DICKSON and SUZANGAR (41) compared the direct incubation and the preincubation methods, using slices of the Yoshida rat sarcoma. With direct incubation in buffer in Warburg flasks, both respiration (gas phase air) and anaerobic glycolysis (gas phase 5% CO_2 in N_2) were completely inhibited after 4 h at 42°C; on transference of the heated cells to fresh medium at 38°C, there was no further gas exchange. When tumour slices were preincubated in Waymouth medium for 4 h at 42°C in a CO_2 incubator and then placed in fresh buffer at 38°C in Warburg flasks, both respiration and glycolysis continued at a low level (approx. 25% of that of non-heated control slices) for a further 4 h before gas exchange ceased. Following preincubation, the tumour slices did not produce tumours in rats, even though a low level of metabolism was still present at the time of inoculation. The question of cells vs. slices for in vitro studies was examined by JARES and WARREN (87). The Brown-Pearce epithelioma of rabbits was incubated in shaking flasks with O_2 as the gas phase. At 42°C, the thermal death time for tumour slices was 30 h compared to 20 h for cell cuspensions, as determined by subsequent transplantability of the heated tumour. With the VX2 rabbit carcinoma, cells and slices exhibited a similar inhibition of metabolism at 42°C under identical incubation conditions (38); the effects of hyperthermia on transplantability were not compared. It is often difficult and time-consuming to obtain adequate numbers of satisfactory cells from tumours (especially human cancers) for testing; tumour slices are usually rapidly obtained and are more convenient than cells for most purposes. Several workers have shown that the inhibitory effects of elevated temperature (42°C) on the synthesis and metabolism of the 45S ribosome precursor RNA and on polysome aggregation are completely reversible after heating periods of up to 3 h (79, 105).

In view of these various considerations, we have now adopted a standard in vitro assay procedure using tumour slices and comprising a 4-h culture (preincubation) period of hyperthermia followed by a 6-h incubation at 38°C; during the latter 6 h respiration and glycolysis are measured in Warburg manometers and concomitant isotope uptake is measured in cultures after renewing the medium. For testing the effect of cytotoxic drugs in combination with hyperthermia, an incubation period of 4 h has been found to give inconsistent results (41). If the tumour is sufficiently active metabolically, then a suitable protocol is to culture slices for 24 h, at 38°C for the initial 20 h and then 4 h at 42°C; metabolic studies are then performed at 38°C over a further 6 h (41).

Glycolysis in cancer cells has proved to be less generally affected by elevated temperature than respiration. In some tumours inhibition of O_2 uptake is accompanied by inhibition of glycolysis (39-41), while in others (19, 119, 122) glycolysis is unaffected by hyperthermia. The latter cases appear to represent an "uncoupling" of these usually inter-related parameters (39, 41) to the detriment of the cells' replicative ability, since such glycolyzing cells fail to produce tumours on inoculation into host animals (39, 122). DICKSON and SUZANGAR followed the changes that occured in respiration and glycolysis of Yoshida tumours regressing over 12-14 days after heat treatment (41). In tumour material examined immediately after hyperthermia, O_2 uptake and CO_2 production were both reduced by approximately 50%. Subsequently, O_2 uptake remained reduced, but CO_2 production recovered to control values by 3 days after treatment. In further metabolic studies, Yoshida sarcomas were subjected to a heating period that was inadequate to cure the tumour (36). Following hyperthermia, anaerobic glycosis was significantly reduced while respiration was unaltered, and there was enhanced tumour cell dissemination in the rats. These findings lend additional support to the postulate of MONDOVI et al. (119) that the damaging effect of heat on tumour cell respiration should be regarded as a general phenomenon. The case for predicting tumour sensitivity to hyperthermia from an irreversible decrease in respiration of the tumour heated in vitro would also appear to be strengthened. In the work of DICKSON and SUZANGAR (41) the 50% irreversible reduction in O_2 uptake was accompanied by a progressive decrease in nucleic acid and protein synthesis and complete tumour regression.

The question of the significance of a less than 100% inhibition of O_2 uptake obtained on in vitro heating in relation to tumour sensitivity in vivo cannot be answered at the present time. The situation occurs most commonly with human tumours. In this case, a 30% inhibition or respiration and/or glycolysis may be accompanied by a high inhibition of precursor uptake into DNA, RNA and protein following a 4-h preincubation of tumour clices at 42°C (40). Such heated tumour tissue does not grow in immune-suppressed rats, mice of hamsters while 50% of control unheated tumours do grow under these conditions. Comparative studies on the in vitro response of human tumours and the response of the same tumour in heated patients have not yet been accomplished. The relationship of a partial and irreversible inhibition of respiration to the biological malignancy of a tumour remains undetermined, therefore. Evidence is accumulating for the biochemical (16, 137) and biological (16, 32, 99) diversity of solid tumours, and current views on tumour cell kinetics emphasize the therapeutic importance of the non-proliferating fraction of cells in a tumour population (56, 161). Non-proliferating cells are less susceptible to damage by heat than multiplying cells (38, 39). A partial inhibition of metabolism in vitro after hyperthermia may therefore reflect only the response of the growth fraction of the tumour, in which case a longer heating period may lead to greater inhibition. As discussed later, cells that are not destroyed by heat recover rapidly and fully. It must also be remembered that the malignant cell can avail itself of numerous metabolic pathways, and inhibition of one pathway may simply lead to more efficient utilization of another (as, for example, in the Pasteur and Crabtree effects). Irrespective of alterations in the inter-related pathways of respiration and glycolysis, a cell cannot replicate in the absence of DNA, RNA and protein synthesis. The problem arises when inhibition of these parameters in vitro by heat is only partial and extrapolation to the in vivo situation is hazardous. The importance of measuring both respiration and glycolysis and associated changes in precursor uptake into DNA, RNA and protein in predictive in vitro systems for hyperthermia sensitivity testing cannot be over-emphasized; if heat causes inhibition of these parameters, the inhibition must be examined for irreversibility. The need for in vitro-in vivo assay systems to evaluate the results of Warburg manometry and of biochemical studies, in terms of growth potential of the cells being examined is evident, and has been stressed in publications from this laboratory (39, 41).

b. Techniques

For culture or manometry, tumour is removed from the host animal under sterile conditions, placed on ice, cut into pieces, and washed with Rinaldini saline (141) containing penicillin (100 μ/ml), streptomycin (100 μg/ml), and mycostatin (100 μ/ml). If cells are being used, the tumour pieces are then disaggregated as rapidly as possible by enzymes such as trypsin, DNase, collagenase (see e.g. ref. 122). The suspension is filtered through stainless steel sieves (80 mesh followed by 200 mesh) and the cells suspended in culture medium (we routinely use Waymouth MB 752/1 medium supplemented by 10% pooled human AB serum). The cells are counted on a haemocytometer and assessed for viability using trypan blue dye (84). Cells should not be used for transplantation or metabolic studies if the population viability is less than 80%. For studies on tumour slices, the material is cut into thin fragments (less than 1 mm^3) with cataract knives or conveniently shaped surgical blades, and kept at 4^oC until required.

Oxygen consumption (respiration) and CO_2 production (glycolysis) are measured by conventional Warburg manometry, using 5-10 million tumour cells or 50-200 mg tumour slices per flask. Incubations are carried out simultaneously at 37.5^oC or 38^oC (depending on the body temperature of the host animal) and at the required elevated temperature (e.g. 42^oC). Respiration is studied in a Krebs-Ringer phosphate (KRP) buffer, pH 7.4, containing sodium succinate (0.013 M), with 0.2 ml 10% KOH in the centre well and air as the gas phase. Anaerobic glycolysis is measured in a Krebs-Ringer-bicarbonate-phosphate (KRBP) solution (119), pH 7.4, containing glucose (2 g/l) and a gas phase of 95% N_2/5% CO_2 (initial O_2 content of mixture less than 20 ppm, Air Products Ltd., Gateshead, England). For all tumour types we have examined (animal and human), the KRBP buffer maintains a more stable pH over the incubation period than the traditional KRB buffer; the additional phosphate is especially useful for stabilizing the pH with highly glycolytic tumours. A stable incubation buffer pH is required in hyperthermia experiments, since it has been claimed that the deleterious effects of heat are potentiated by low pH (4).

Isotope Studies. For isotope uptake studies, radioactive thymidine, uridine or leucine are added to cultures of tumour slices at an initial concentration of 0.2 μCi/ml medium. Following culture, the tissue is extracted by a modified Schneider procedure, and activity in the samples expressed as cpm/ug DNA phosphorus (41).

4. THE EFFECTS OF HEAT ON ANIMAL TUMOURS

A. Biochemical Changes

Figure 1 records the O_2 uptake values for VX2 carcinoma cells incubated at temperatures above 37.5^oC. O_2 uptake increased with elevated temperature up to 40^oC. At 41^oC, O_2 uptake was similar to that at 37.5^oC, while at 41.5^oC an inhibitory effect on respiration became apparent after 5 h. At 42^oC, there was a significant depression of respiration as early as 60 min, compared to the control cells at 37.5^oC ($P < 0.001$). After 2 h at 42^oC, respiration was totally inhibited; the heated cells

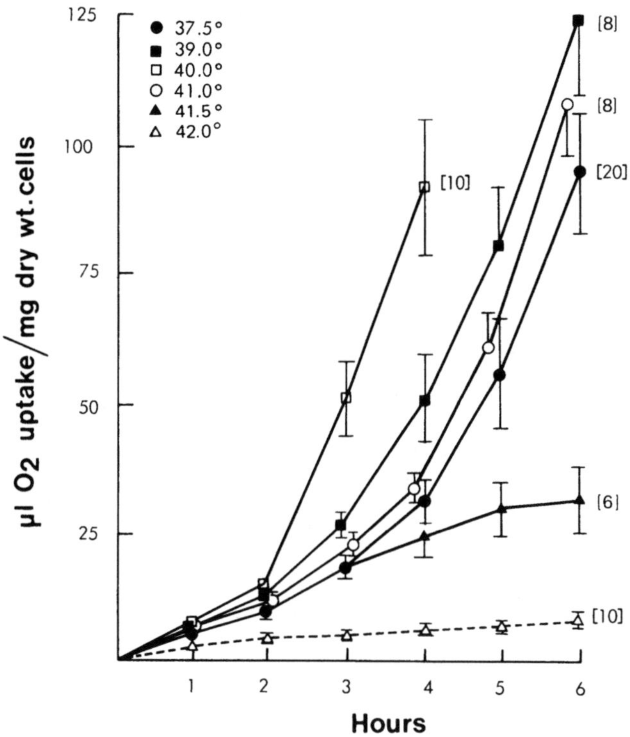

Fig. 1. Oxygen uptake of VX2 cells (5 - 10 x 10^6) with time in Tris-HCl-sucrose buffer at temperatures above 37.5°C. Each point is the mean ± S.D. for the number of tumours indicated in brackets

ceased respiring after 2 h when returned to 37.5°C and on inoculation into rabbits did not produce tumours, i.e. the inhibition of respiration was irreversible and corresponded with annulment of the biological malignancy of the cells. At temperatures above 42°C, the inhibition of respiration was more rapid -- at 43°C, for example, O_2 consumption was reduced to zero after 1 h. We have confirmed this picture of an increase in O_2 uptake up to 40°C and a decrease at 41.5°C and over with the two rat tumours, the SDB tumour (39) and the Yoshida sarcoma (41).

Several workers have shown that the destructive effects of hyperthermia on susceptible tumours depend upon both the degree of heat applied and its duration. The minimum time required to cause destruction of a tumour at a given temperature is referred to as the thermal death time. The response of the neoplasm can be expressed as an exponential type of curve in which approximately the same biological effect is obtained by halving the exposure time for each degree that the temperature increases above 42°C. WESTERMARK (183) measured the thermal death times in the Flexner-Jobling carcinoma and Jensen sarcoma from the times required to produce cessation of O_2 uptake and CO_2 production in Warburg flasks. For both tumours, these times ranged from 90 min at 45°C, through 30 min at 47°C to 13 min at 49°C; the heated tumour slices were not transplanted into rats. The times required to produce tumour regression in vivo were similar to the corresponding times required to produce inhibition of metabolism in vitro. JOHNSON (90) compared the in vitro and in vivo thermal death times for the Walker 256 carcinoma and Jensen sarcoma. His results compared favourably with those of WESTERMARK, but he found that longer times were required _in vivo_ than _in vitro_ for tumour destruction, as assessed by the ability of the tumour heated in vitro to grow on transplantation into rats. The thermal death times _in vitro_, as determined by capacity for growth on transplanting were: 47°C, less than 20 min; 45°C, 1 h; 43.5°C, 3 h. After 3 h exposure at 42.5°C, 50% of the tumour samples took on transplantation, but a similar exposure at 41.5°C had no damaging effect. For slices of mouse sarcoma 180, the Walker 256 carcinoma and the Jensen sarcoma of the rat, JARES and WARREN (_87_) found that at 42°C, 8 h, 7 h and 4 h respectively were required to irreparably damage the tissue; at 41.5°C there was a significant increase in the thermal death times of these tumours to 15 h, 13 h and 10 h respectively. In more recent years, these findings on the relationship between exposure time to heat and tumour destruction have been corroborated by experiments _in vivo_ on sarcoma 180 (_30_) and a mammary carcinoma (_127_) in mice. In 1927, WESTERMARK reviewed the data up to that time on the heat tolerance of normal tissues. Although the older investigations were at fault because the temperature of the surroundings but not that in the heated tissue was measured, the results of various workers with different tissues and organs indicated that normal tissues withstood higher degrees of heat than tumours. In the few cases in which individual tissues were studied at different temperatures, the thermal death times followed a similar exponential type of curve to that for cancer tissues with an important qualification illustrated by the work of FUKUI (_54_, _55_) on the thermal dose required to produce degeneration of the germinal epithelium of the testis. The results were: at 48°C, 20 min; 47°C, 30 min; 46°C, 1 h; 45°C, 2.5 h; 44°C, 4,5 h. These values from 48°C to 44°C fall on a straight line when graphed on semi-log paper. At temperatures below 44°C, the values deviate significantly from an exponential relationship and much longer times are required than expected. At 41°C, for example, the thermal death time obtained was 100 h, 3 times longer than that predicted by the exponential relationship. WESTERMARK confirmed these findings for the pattern of thermal death times of germinal epithelium and liver in rats, and by direct comparative studies with Jensen sarcoma and Flexner-Jobling carcinoma showed that the differences in heat sensitivity between the tumours and the normal tissues were greatest at the lower temperatures examined, 44°C in this case. The differential heat sensitivity became reduced at higher temperatures and was not perceptible at 48°C in WESTERMARK's work. More recently, CHEN and HEIDELBERGER (_21_) studied the thermal sensitivity of a cell line that originated from C3H mouse prostate and of chemically induced malignant cells derived from the line. After an exposure of 90 min at 43-44°C, the transformed cells had a greater reduction in percentage survival than the normal cells; at 45°C there was no difference in survival between the two cell types. GIOVANELLA et al. (_59_) have reported that the differential heat sensitivity between fibroblasts and sarcoma cells derived from them is no longer present at temperatures as low as 43°C. We have subjected many normal rabbit and rat tissues, such as lung, liver, kidney, skin and blood to 42°C -- in all cases respiration and glycolysis have shown no significant depression at the elevated temperature. As discussed later, temperatures above 42°C are not feasible for the treatment of tumours in animals, either by local or total body hperthermia. Similarly, in humans, total body hyperthermia at temperatures in excess of 42°C leads to problems of host tolerance such as cardiac arrhythmias and evidence of cerebral disturbance, and with local hyperthermia, severe complications occur following perfusion at such temperatures (_19_).

On the basis of these various considerations, therefore, we have proposed that for practical purposes the lower end of the thermal death range for susceptible tumours begins at 41.5°C, and for therapy the operational range for animals and for man is 41.5-42°C (35, 38, 41). There is considerable evidence, as detailed above, that at temperatures in excess of 42°C the differential effects of heat on cancer cells diminish, whilst a small reduction in temperature from 42°C to 41.5°C produces a marked decrease in the destructive effects of heat. Consequently, in our own work we have adopted an operating intra-tumour temperature of 42°C for hyperthermia studies; this temperature represents a good compromise between host tolerance considerations and an unduly prolonged heating time.

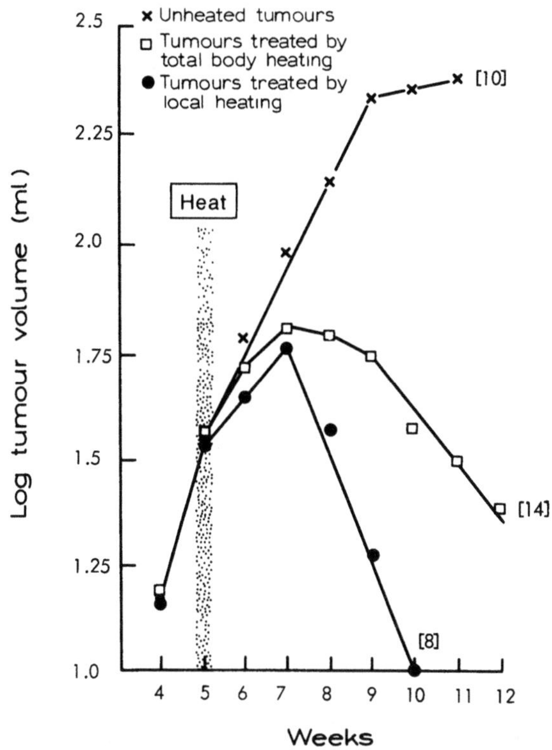

Fig. 2. Growth curve for VX2 carcinoma in rabbit following inoculation of 1 x 10^6 tumour cells into rear limb muscles, and changes in tumour volume following local and total body hyperthermia applied in fractionated doses of 1 h each on days 35, 36 and 37 after tumour implantation

B. Histological Changes

Figure 2 shows the growth curve obtained for the VX2 carcinoma following inoculation of 1 x 10^6 cells into the muscles of the rabbit's hind limb. The tumour became palpable by approximately 4 weeks after inoculation and by 5 weeks the volume was about 40 ml. From 5 to 9 weeks, tumour volume increased exponentially and then the rate of increase diminished in accordance with the "tail" portion of the Gompertz growth function for solid tumours (98). Untreated rabbits died at 70 $\pm$ 6 days after tumour cell inoculation.

Heating, local or total body, was applied at 35 days at the beginning of logarithmic growth (Fig. 2). Following a single application of heat to the VX2 carcinoma (intra-tumour temperature 42°C for 1 h), there was congestion of the small blood vessels so that the tumour was oedematous and deep red in colour when sectioned (122). Similar changes have been described previously for mouse (126, 144) and rat (90, 183) tumours following heating, and in the VX2 there was no ensuing permanent cell damage after one or two such heatings separated by a 24 h interval. Permanent alterations in the tumour were not observed with a heating period less than 3 h. As the rabbits did not tolerate such prolonged heating, the hyperthermia was fractionated into three 1 hour periods on consecutive days (days 35, 36, 37 after tumour inoculation).

Following this regimen of heat treatment, there were rapid and striking histological changes in the tumours (Fig. 3, ref. 122). In the untreated tumour, the cells formed sheets and irregular masses, separated by a fine connective tissue stroma (Fig. 3A). Twenty-four hours after the third heat application, the tumour showed widespread necrosis, with pyknosis, karyorrhexis and cell lysis (Fig. 3B and C). Intense fibroblast and macrophage activity then occurred (Fig. 3D and E) with subsequent replacement of the tumour architecture by connective tissue in which no cancer cells could be identified (Fig. 3F).

Although numerous workers have heated tumours in animals by various techniques, comparatively little attention has been devoted to microscopic examination of the tumours following treatment. ROHDENBURG and PRIME (144) carried out detailed histological analysis on Walker 256 carcinomas and Jensen sarcomas heated by high frequency currents. Twenty-four hours after treatment, the tumours were the seat of intense hyperaemia but no changes were observed in the tumour cells. Three days after heating the cell boundaries began to disappear, although the nuclei retained their outlines and staining capacity; the connective tissue stroma was clearly observed as were the intact blood vessels. Subsequently, the gradual death of the tumour cells continued and the tumour was replaced by fibrosis extending from the persisting stroma and blood vessels. The results of WESTERMARK's histological studies (183) on the Flexner-Jobling and Jensen tumours were in accord with those of ROHDENBURG and PRIME. Morphological changes were not observed in the tumour cells until a few days after heating; necrotic changes became more prominent with time until by 6 days the tumours showed complete necrosis. Further details of regression are not given, except to state that the tumours were resorbed and not sloughed off after hyperthermia. Likewise, JOHNSON (90) did not particularize the alterations occuring in Walker 256 carcinomas and Jensen sarcomas receding after short-wave diathermy, except to record that from 3 to 10 days after heating there were large areas of necrosis with cells in which the nuclei were pyknotic and the cytoplasm stained poorly; leucocytic infiltration was rarely marked. JOHNSON commented that the histological appearances were quite distinct from those characteristic of tumours irradiated by X-rays or gamma-rays.

OVERGAARD and OVERGAARD (127) have recently reported detailed histological and histochemical studies on subcutaneously transplanted mammary tumours in inbred mice heated by local short-wave diathermy. Changes in the cells following heating included clumping of the nuclear chromatin, vacuolation of the cytoplasm with damage to the mitochondria, arrest of mitosis and then pyknosis. The rapidity of onset and the intensity of the changes increased with the dose of heat used. The findings were similar to those reported over 30 years previously by OVERGAARD and OKKELS (126) for the Crocker mouse sarcoma 180 heated by diathermy to 42-46°C. OVERGAARD and OVERGAARD found that the destructive effects of heat were confined to the tumour cells; the normal cells of the connective tissue stroma and of the tumour blood vessels were undamaged. It was concluded that the histological appearance after heat treatment is specific and differs from the usual necrotic and necrobiotic processes seen in tumour tissue -- the latter involve both the normal and the malignant components of the tumour mass. A "moderate macrophage reaction" was seen in the heated tumours and a very rapid resorption of dead tumour cells, with a concomitant rapid new growth of connective tissue, were also stated to be characteristic of the reaction. The rapid nature of the resorption process may have been connected with the fact that the tumour studied by OVERGAARD and OVERGAARD was a syngeneic one; immunological reactions probably play only a minor part in the response of these small mammary tumours (maximum volume < 0.2 ml) that do not metastasize. DICKSON and SUZANGAR (41) compared the histology and concomitant biochemical events in allogeneic solid Yoshida tumours regressing after heat or drug treatment. Following 2.5 mg methylenedimethane sulphonate (MDMS), tumours (1.5 ml volume) disappeared within 8-10 days. There was rapid tumour cell pyknosis and fibroblast infiltration accompanied by a progressive decrease in respiration, glycolysis and labelled precursor uptake into DNA, RNA and protein in the regressing tumour. Hyperthermia (intra-tumour temperature 42°C for 1 h) resulted in a significantly less rapid regression rate with tumour disappearance in 12-14 days. Cell death was less rapid than after MDMS, there was a slower removal of dying cells and a more gradual replacement of tumour architecture by connective tissue. This histological picture was accompanied by a protracted decrease in isotope uptake and an inhibition of respiration. When heat and drug treatment were combined, the effect of the heat was to significantly slow down the rate of tumour regression. The rats had regional lymph nodes that were enlarged and contained tumour cells at time of treatment; following drug or heat therapy the nodes returned to normal and the survival time of the animals was similar to that of control untreated rats (41).

It is of note that most workers who have heated tumours in animals have not achieved a regression rate of 100% in the treated tumours; partial regressions followed by progressive growth and also local recurrence of 'cured' tumours have occurred. With electrical methods of heating, e.g. short-wave diathermy, it has been found that even with the most refined apparatus, the heat produced is not uniform throughout the tumour (90, 127, 183); if intra-tumour temperature has not been measured,as e.g. in CRILE's work on the treatment of mouse and dog tumours by water bath immersion (29, 30), it cannot be assumed that all malignant cells have attained an adequate temperature for destruction. In some work, the size of the treated tumours has varied considerably, e.g. between 0.5 ml and 2.0 ml (90), between 0.3 and 5.8 ml (183). JOHNSON (90) found that with the tumours and volumes examined (Walker 256 carcinoma and Jensen sarcoma up to 2.0 ml) the response to heating was independent of tumour size. With the Jensen sarcoma and Flexner-Jobling carcinoma, WESTERMARK stated that between the limits of 0.3-2.4 ml the effects of hyperthermia were not related to tumour mass. It has been found that with the solid Yoshida sarcoma there is a relationship between tumour volume and curability by heat (37); beyond the exponential phase of tumour growth it requires an increased amount of heat to destroy the tumour (see later). This is in keeping with current views that as a tumour increases in size there is a decrease in the number of proliferating cells (53, 56, 161); hyperthermia is most effective against multiplying cells (see 19 for refs.) and non-proliferating cells require a longer heating period for destruction (39). It must be mentioned that with sarcoma 180 in mice, CRILE claimed the tumour became more heat-sensitive as it grew larger (30). As the data was not defined in terms

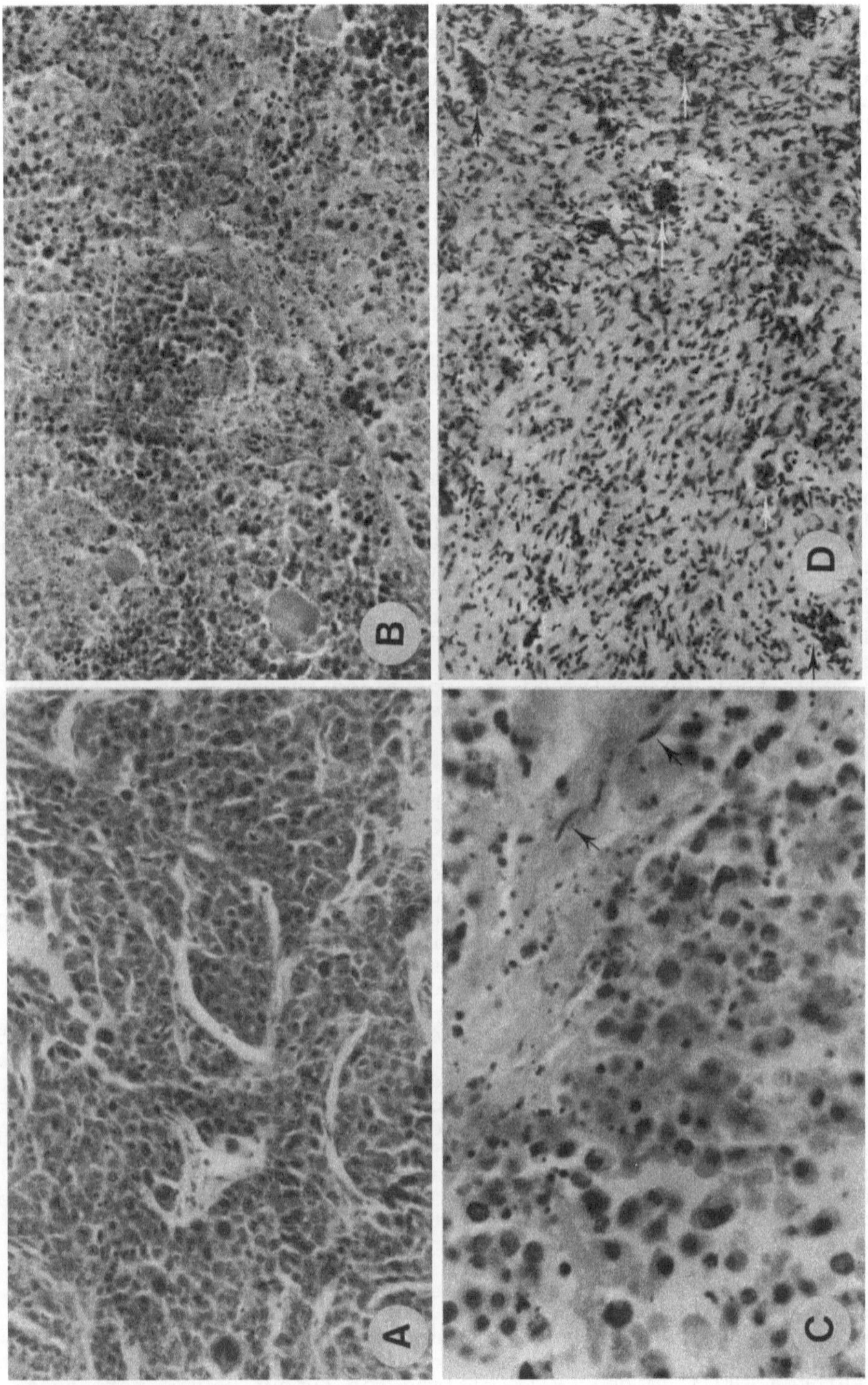
B
D
A
C

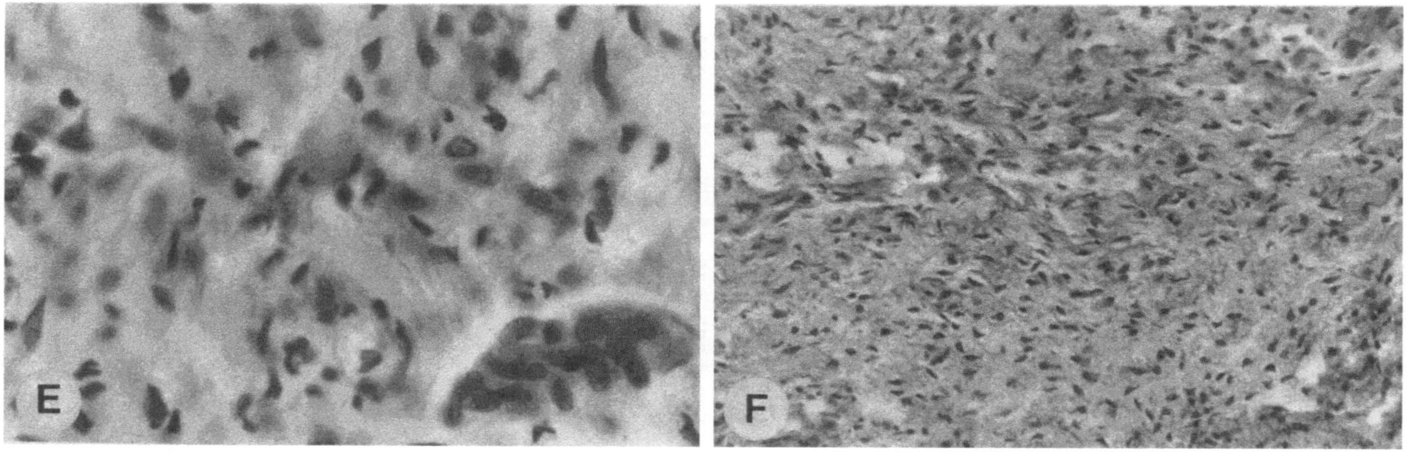

Fig. 3. Histology of VX2 carcinoma following curative hyperthermia.
A. Untreated tumour which consists of sheets and islands of cells in a fine connective tissue stroma. x 200.
B and C. VX2 tumour 24 h after third application of hyperthermia. There is widespread necrosis with pyknosis and karyorrhexis. In C a few normal-appearing fibroblasts of connective tissue that has survived heating are seen (arrows). B x 200; C x 700.
D and E. Intense fibroblast and macrophage activity replacing dead tumour at 4 weeks after heat treatment. Groups of macrophages are shown by arrows in D, and one group of these cells at higher magnification in E. D x 200; E x 700.
F. VX2 carcinoma 12 weeks after hyperthermia. Tumour architecture has been replaced by fibrous tissue and no cancer cells can be seen. x 200

of a tumour growth curve, its interpretation is uncertain. In their series of human limb tumours treated by regional hyperthermic perfusion, CAVALIERE et al. observed that the histological effects were sometimes immediate, while in other cases they were delayed (19). Complete regression of the tumour with non-recurrence and a good prognosis was correlated with total necrosis of the tumour with no viable cells visible in any of the sections after treatment.

The above review of data emphasizes the importance of standardizing the tumour system and the heating technique in hyperthermia experiments if meaningful information is to be obtained. Variables such as tumour volume and inhomogeneous heating must be controlled and a wider use of detailed and more sophisticated histological (e.g. staining for macrophages and plasma cells, histochemical studies of enzymes) and immunological (e.g. determination of changes in tumour cell antigenicity) methods is required before we can decide whether and to what extent the response to heating is characteristic and specific and involves the immune system. Species differences in the host must be kept in mind with these investigations. It does seem established, however, that tumour regression without recurrence requires destruction of 100% of the tumour cell population. This reaffirms the concept of total cell kill advocated for eradication of cancer by SKIPPER et al. from studies on tumour response to cytotoxic drugs (162), and emphasizes the danger to the host of cells that survive therapy, even though such cells are in a seemingly dormant state (Go cells, see 56).

C. Local Hyperthermia vs. Total Body Hyperthermia

The changes in tumour volume following three heatings at 24 h intervals (local hyperthermia or total body hyperthermia) are illustrated in Figure 2. Tumour volume continued to increase until the 7th week, after which all heated tumours decreased in volume. In Figure 4, mathematical regression lines have been fitted to the log tumour volume measurements following heating. The difference in slope of the regression lines is statistically highly significant ($P < 0.001$). In terms of volume, the regression equations mean that, in the case of local heating, tumour volume was halved every 0.97 week from the 7th week after inoculation, while after total body heating, tumour volume was reduced by half every 2.9 weeks over the same period (for a detailed statistical treatment of this data see ref. 38).

In terms of animal survival, 4 of 8 rabbits treated by limb immersion are alive, with no signs of tumour, almost 3 years after treatment. Of 14 animals treated by total body heating, 4 died at 20 weeks after tumour inoculation, 1 survived to 30 weeks and 1 is alive 2 years after therapy; the other 8 rabbits died within 10 weeks of tumour inoculation (mean survival time 65 $\pm$ 8 days). At autopsy, all animals that died had metastases in the lungs and inguinal and para-aortic lymph nodes.

What is the explanation of the difference in survival following the two methods of heating? Figure 5 compares the intra-tumour and body temperature recordings over the period of hyperthermia in the two methods. In each case, the intra-tumour temperature rose to the region of 42°C within 15-30 min of applying the heat. Over the 1 h heating period between 15 and 75 min (bath temperature 46°C) the mean intra-tumour temperature was 42.7°C for local hyperthermia and 42.6°C for total body heating, and the temperature distribution patterns in the tumours were similar in the two methods. With water bath immersion, the rectal temperature of the rabbits increased to the region of 40°C and the animals were able to cope with the increased heat load by an elevation

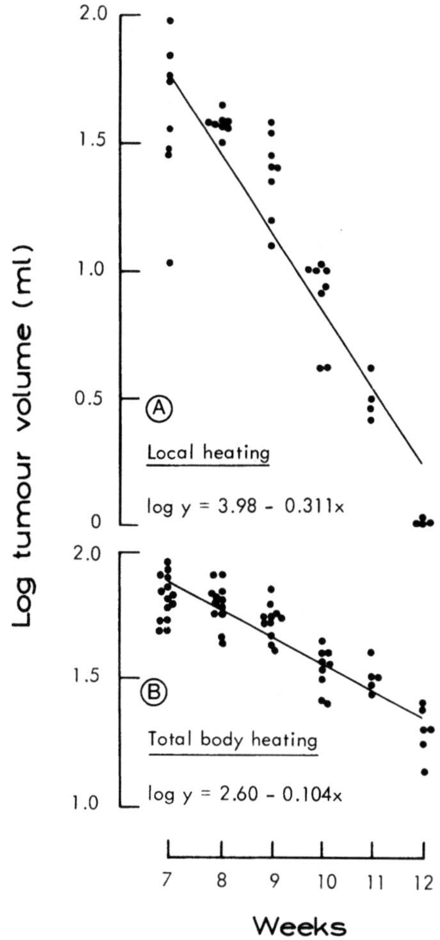

Fig. 4. Regression lines fitted to individual log tumour volume measurements following local and total body heating and illustrated graphically in Fig. 2

of skin temperature (Fig. 5A). In total body heating at an environmental box temperature of 46°C, the central body temperature as measured in the rectum averaged 42.2°C over the 60 min heating period (Fig. 5B). However, when intra-thoracic and intra-abdominal heating patterns were examined by means of sensors placed below the heart and below the liver, it was found that these temperatures fluctuated. Over the heating term (15-75 min) the temperatures oscillated between a minimum of 43.6°C and a minimum of 39.8°C. The pattern of fluctuation varied between individual animals, but over the therapy period the average intra-thoracic temperature was 40.5 ± 1.0°C and the average intra-abdominal temperature was 40.8 ± 0.7°C.

In the rabbit VX2 system metastasis is rapid and predictable, and by 7 days after incoculation tumour is present in the regional lymph nodes as previously described by EDWARDS (44). When the tumours are heated at 5 weeks the rabbits also have secondary tumours in the para-aortic lymph nodes and lungs (Fig. 6). We have postulated that following local heating, tumour cell death and the absorption of necrotic material in the primary tumour mass occur in a characteristic manner that may lead to stimulation of the host's immune system, thus enabling the host to destroy metastatic cells (Fig. 7 and refs. 35, 38, 41, 122,

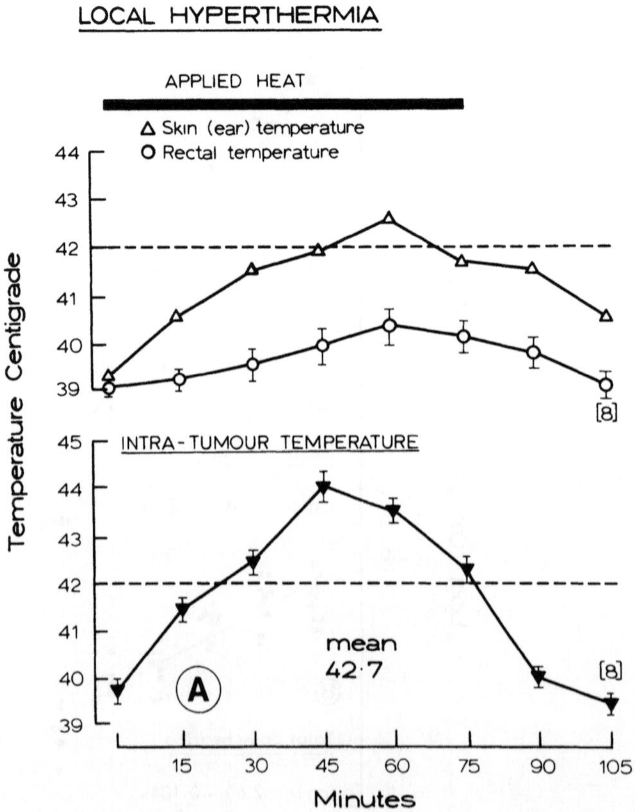

Fig. 5A. Profiles for intra-tumour and body temperature during treatment of VX2 carcinoma in rabbit by local hyperthermia. Patterns of ear, intra-abdominal and intra-thoracic temperature during heating varied from rabbit to rabbit; temperature response curves for one animal are shown

123). The presence of large numbers of macrophages in the resolving tumour mass (Fig. 3D and E) has been consistently noted to be marked and prolonged (122). Macrophages are now regarded as forming a third component of the immune system (76) and the co-operative functions of these cells in relation to both the cell-mediated and serum-mediated (antibody) components of the immune response are now well recognised (146, 182). MONDOVI et al. (118) recently described an increased antigenicity in Ehrlich ascites cells heated at 42.5°C in vitro for 1-3 h, and antigens associated with macrophages have been reported by HUMPHREY (86) to be more immunogenic than the native antigen. In the present work, animals challended with 1 x 10^6 viable VX2 cells into the contralateral leg 3 months after local heating did not develop tumours. The concept of stimulation of the immune system by tumour products as a result of heat therapy is reminiscent of the early experiments of FOLEY (50), who immunized animals to methylcholanthrene-induced tumours by ligating the tumours and allowing them to resorb. Similarly, LEWIS and APTEKMAN (100) induced immunity in rats bearing sarcomas or carcinomas by occluding the tumour blood supply with a tourniquet. Immunity developed even when the tumour was excised shortly after it was strangulated. If the tourniquet was applied too tightly, this prevented absorption of the strangulated tumour and interfered with the production

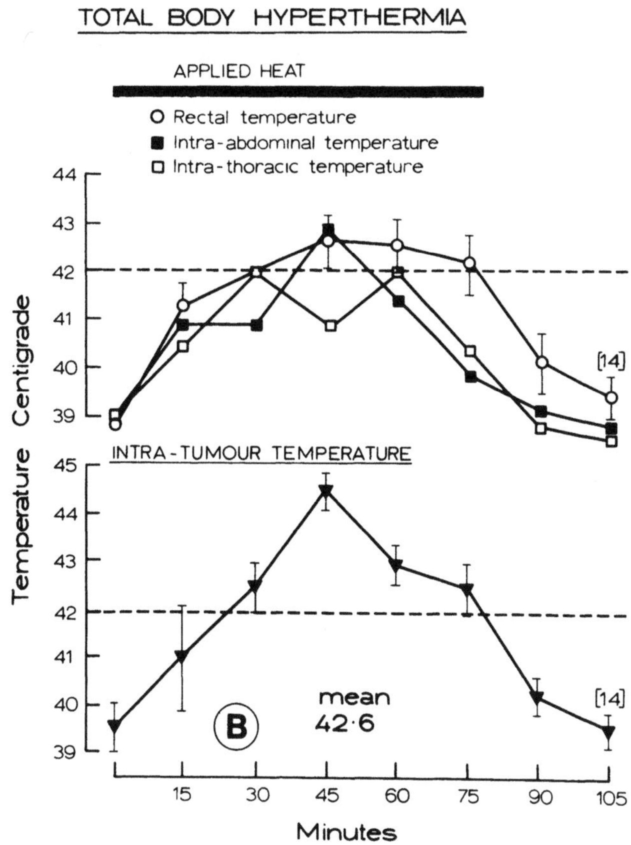

Fig. 5B. Profiles for intra-tumour and body temperature during treatment of VX2 carcinoma in rabbit by total body hyperthermia. Patterns of ear, intra-abdominal and intra-thoracic temperature during heating varied from rabbit to rabbit; temperature response curves for one animal are shown

of resistance in the host to subsequent tumour implantation. Further experiments by APTEKMAN, LEWIS and KING in the early 1950s (see 133 for refs.) indicated that the optimum conditions for inducing tumour immunity in rats were: (1) that the tumour must actually grow in the host (2) that it must die in the host (3) that it must degenerate under conditions which permit relatively slow absorption of degeneration products. As discussed later, there is additional strong circumstantial evidence that an active immune response may be evoked by local hyperthermia in man and in animals (see p. 85). That 54% of the rabbits still died of metastases after local heating may indicate that in these cases the residual host burden of cancer cells was too large to be overwhelmed by the immune system (109).

The poor results with total body heating in the rabbit could be due to several factors. At temperatures above 37.5°C the metabolism of the VX2 cells increased, as would be expected on kinetic grounds (Fig. 1); at 40°C, O_2 uptake was treble the control value and inhibition did not become evident until 41.5°C. It is possible, therefore, that the oscillating intra-thoracic (average 40.5°C) and intra-abdominal (average

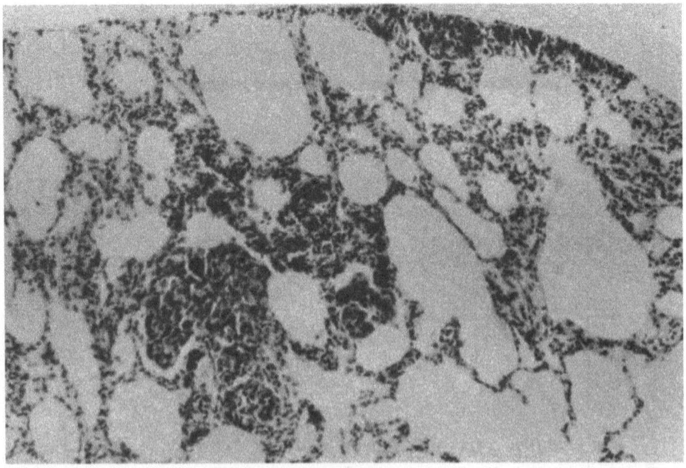

Fig. 6. Section of rabbit lung showing metastatic VX2 deposits in parenchyma and on pleural surface. x 150

40.8°C) temperatures were inadequate to destroy the metastatic cells. Inability to maintain the temperature above 41°C may have contributed to the death of the animals by stimulating activity of the secondaries (Fig. 7). Secondly, in total body heating, it is tacitly assumed that

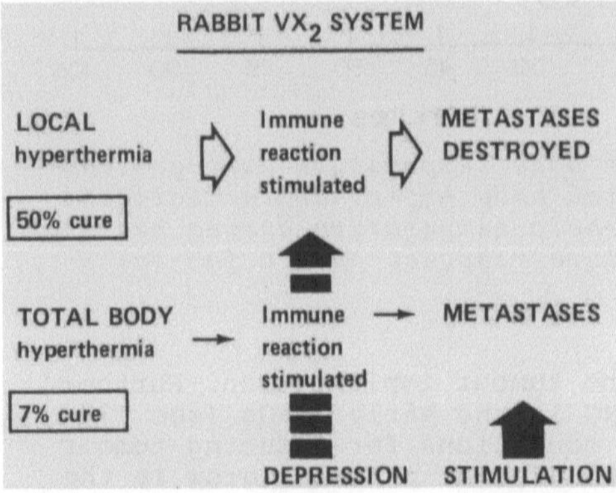

Fig. 7. Scheme to illustrate the difference in results obtained in treatment of VX2 carcinoma by local heating and total body heating in terms of postulated effects on host immune system

metastatic and primary tumour cells are equally sensitive to heat. Table 1 shows that for the VX2 tumour this is not the case. Although respiration at 37.5°C was comparably in both types of cell, the effect of heat was less marked on the metastatic cells than on the primary tumour cells - whereas O_2 uptake was inhibited rapidly in the primary cells, the secondary cells continued to respire at reduced but linear rate at 42°C. Thirdly, it may be that total body heating depresses the immune response (Fig. 7). As the primary tumour is maintained at a similar elevated temperature in both types of

Table 1. O_2 uptake of primary and metastatic VX2 cells at 37.5°C and 42°C

	O_2 uptake (µl/mg, dry wt)			
	1 h	2 h	3 h	4 h
37.5°C				
VX2 cells	6.3 ± 1.0	10.5 ± 1.9	18.9 ± 2.5	31.0 ± 3.2
Metastatic VX2 cells	6.9 ± 0.6	13.6 ± 1.0	16.8 ± 1.8	22.3 ± 2.4
42.0°C				
VX2 cells	3.9 ± 0.7	4.7 ± 0.7	5.1 ± 0.9	5.7 ± 0.8
Metastatic VX2 cells	3.7 ± 0.3	8.8 ± 0.3	11.4 ± 0.7	14.2 ± 0.8

The Warburg flasks contained 5-10 x 10^6VX2 cells from the primary limb tumour or from the groin lymph nodes draining the tumour area. The metastatic cells were obtained at 5-6 weeks after tumour inoculation when the nodes were grossly enlarged and almost totally replaced by tumour. The values are the means ± S.D. for 10 tumours and the corresponding regional lymph nodes.

heating, it is assumed that the immune system would be similarly stimulated in each case by altered immunogenicity of the cancer cells and/or by tumour breakdown products. The histological changes in the primary tumour are similar following local and total body hyperthermia, but tumour regression is significantly slower after total body heating (Fig. 4). This may indicate that although in total body heating the stimulus to the immune system from the primary tumour is present, the immune response is less efficient at elevated temperature; or, the immune response may be inadequate in the presence of increased activity of the metastatic cells. The above considerations imply that an immune response is involved in the regression of both primary and secondary tumours in the VX2 system. A further point in total body hyperthermia that also comprises the response of the host is the effect of elevated temperature on normal tissues. This is discussed later.

5. HAZARDS AND POTENTIATORS OF HYPERTHERMIA IN THE TREATMENT OF CANCER

A. Hazards of Hyperthermia

Recent research has disclosed that the use of elevated temperature to treat cancer may involve hazards to the host. This aspect of hyperthermia has emerged from studies on the inadequate application of heat to the solid Yoshida sarcoma in rats (35, 36).

After implantation into the muscle of the hind leg in rats, the Yoshida tumour spread directly into the muscles of the pelvis and abdomen, and metastasized to the gut and heart and to lymph nodes throughout the abdomen and thorax. The tumour was sensitive to hyperthermia *in vitro* as illustrated in Figure 8. Both O_2 uptake and CO_2 production of Yoshida slices were inhibited after 4 h at 42°C, and the inhibition was irreversible as shown by cessation of gas exchange on transfer of the heated tumour to fresh medium at 38°C and failure of the treated slices to take on transplantation into rats (41).

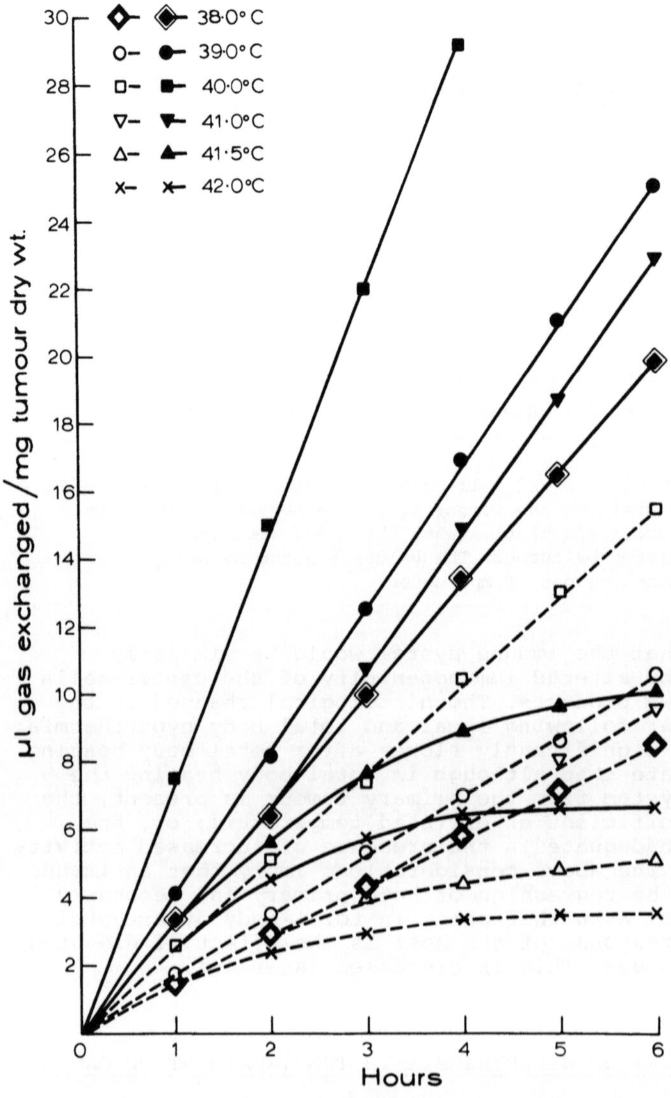

Fig. 8. O_2 consumption (---) and anaerobic CO_2 production (——) of Yoshida sarcoma slices (50-200 mg) in Warburg manometers at temperatures above 38°C. For respiration, buffer was Krebs-Ringer phosphate, pH 7.4, and contained 0.013M sodium succinate; for anaerobic glycolysis, a Krebs-Ringer bicarbonate phosphate buffer, pH 7.4, supplemented with glucose (2 g/l) was used. Each point is mean value from at least 4 experiments, and the S.D.s were 15-20% at all time points

When heat was applied to the leg tumour (volume 10 ml) by water bath immersion (intra-tumour temperature 42°C for 1 h) 8 days after implantation, there was enhanced and widespread dissemination of the cancer (Fig. 9). The animals died at 16 ± 2.5 days after tumour implantation compared with 26 ± 3.1 days in unheated rats, and this difference in mean survival time was highly significant ($P < 0.001$).

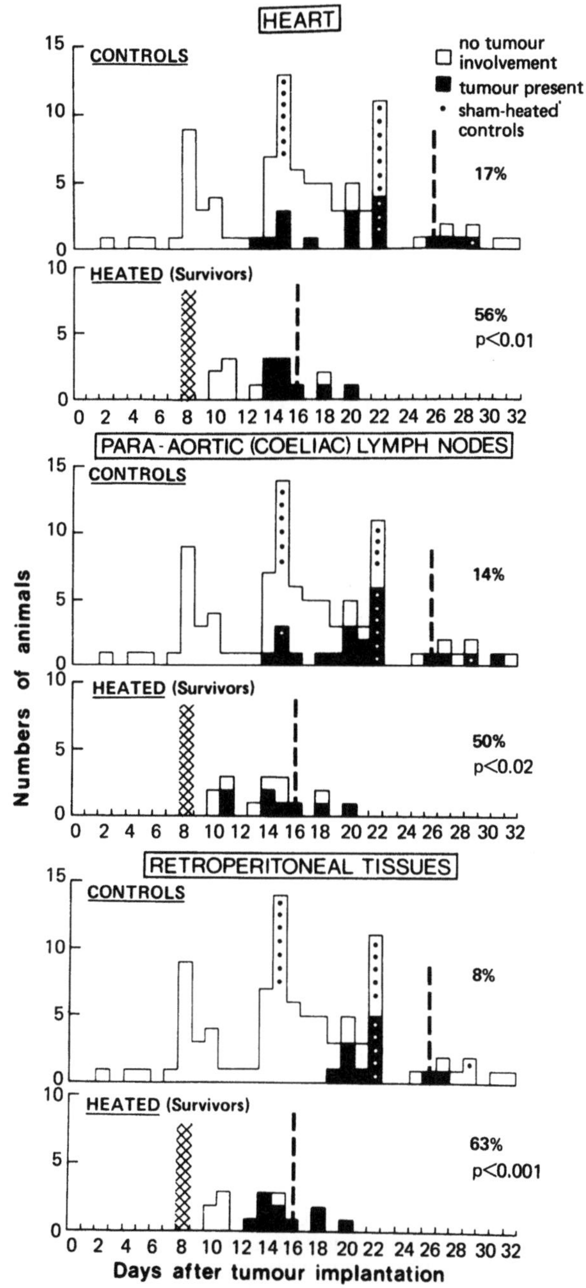

Fig. 9. Frequency of tissue involvement by metastasis following inadequate heating at 42°C of Yoshida sarcoma in hind limb of rats. Profiles illustrate spread by direct, lymphatic and vascular routes. Tumours were heated on 8th day after implantation (vertical hatched line) and percentage figures indicate incidence of involvement of tissues (with significance levels) in heated and control animals between 10 and 20 days. Mean survival time of rats is shown by vertical broken lines. In sham-heated control group , 8 day tumours with needle probes inserted were heated at 38°C for 2 h. At post-mortem, these animals showed a similar pattern of tumour spread as that in the control series

Figure 9 is a detailed analysis of the pathogenesis of the disease in animals that survived the hyperthermia (over 90% of the rats died during or within 24 h after heating. This is discussed later. It must be emphasized, however, that rats which survived recovered rapidly and completely following hyperthermia, i.e. tumour spread was not being studied in chronically ill animals). The tissues chosen illustrate spread of the Yoshida sarcoma via the direct route (into the retroperitoneal tissues), via the lymphatic route (to the para-aortic nodes at the level of the coeliac artery), and by the vascular route (to the heart). All heated rats died by the 20th day after tumour implantation (12th day after heating) and tumour spread was compared in the heated and control rats between 10 and 20 days after implantation.

In the case of the heart, control and heated animals showed involvement by tumour at the same time (14 days after tumour implantation). However, the incidence of myocardial involvement over the 10-20 day period was significantly higher in the heated group.

The lymph nodes at the level of the coeliac artery and alongside the kidneys were involved earlier in the heated animals, and again the incidence of tumour was significantly higher in the heated than in the control unheated rats.

With the retroperitoneal tissues, there was again a premature infiltration in the heated animals, and a significantly higher incidence of tumour involvement over the 10-20 day period.

A further distinction between control and heated rats was observed in lung and liver involvement by tumour (36). Pulmonary tumour emboli and lung metastases were detected in only 2 of 91 control animals (at 22 and 27 days after implantation, i.e. beyond 20 days). Lung involvement was observed in 3 of 16 heated rats, with tumour emboli in 2 of these -- a much higher incidence than in the controls. Liver metastases were not detected in any of the unheated animals. In the heated rats, however, tumour infiltration of the portal tracts and sinusoids of the liver was present in 2 of 16 animals at 14 and 15 days respectively after implantation.

Following hyperthermia, there was no restraint of tumour growth and no inhibition of respiration. Anaerobic glycolysis was significantly reduced (60%) immediately after heating but returned to control values within 3 days. The heated tumours were oedematous and dark red in colour and showed little evidence of cell destruction. The reversible biochemical changes, together with the absence of macroscopic or microscopic evidence of damage, indicate that the hyperthermia was inadequate for tumour destruction. The 1 h period of heat therapy not only accelerated the course of this cancer and increased the incidence of metastases, but led to secondaries in an organ not usually the site of tumour involvement.

Figure 10A illustrates the temperature profiles for the tumour, the abdomen and the rectum when the tumour-bearing limb was heated in a water bath at 43°C. As the intra-tumour temperature rose to 42°C and was maintained there for 1 h, the animal's body temperature increased concomitantly to the 41.5-42°C region.

Several possibilities present themselves, therefore, in relation to the accelerated dissemination of this cancer following heating. From the point of view of the primary tumour, the hyperdynamic state of the circulation at elevated temperature, and the local active hyperaemia, may predispose to increased 'showering' of the tumour cells into the blood from the tumour mass (73). It is now recognised, however, that there is no definite correlation between the presence of cancer cells in the blood per se and the development of metastases (51, 181). Local hyperthermia may have altered cell characteristics such as mobility or surface properties. MONDOVI et al. (118) recently described an increased antigenicity of Ehrlich ascites cells heated at 42.5°C in vitro for 1-3 h, and postulated that short exposure to raised temperature caused some modification of the cell surface. From the point of view of the host, the rise in body temperature to more than 41.5°C may have altered the metabolic or biological properties of the tissues so that they offered a more favourable environment for growth of the cancer cells. The importance of the metabolic dependence of metastatic tumour cells on the surrounding normal host tissues in the spread of cancer was first emphasized by PAGET in 1889 as the 'soil' hypothesis (129), and differences in nutritive suitability of the various host tissues as soils for growth is still the most widely accepted explanation for the susceptibility of different organs to metastasis (185). Intimately associated with such changes in the tissues is the increased blood flow, opening-up of arterio-venous anastomoses and increased tissue oxygen tension occasioned by the elevated temperature. A possible depression in the host immune response at elevated temperature (38) leading to more rapid spread of the disease must also be borne in mind. Heat, therefore, may have the capacity to alter both the 'seed' and the 'soil' in the tumour-host relationship, and this is of direct relevance in total body hyperthermia.

Rats tolerate elevated temperature poorly; when body temperatures in excess of 41.5°C are maintained for 1 h the survival rate decreases rapidly, a temperature of just over 41.6°C representing the 50% lethal dose of heat (36). With the leg tumour, it was not possible to increase the duration of heating to 2 h, either as a continous dose or as two 1 h sessions on consecutive days.

To study further the curability of the Yoshida sarcoma, the tumour was implanted on the dorsum of the rat's hind foot. In this site, an intratumour temperature of 42°C could be maintained for up to 2 h by water bath immersion with less severe physiological disturbance to the animal, as indicated by the smaller increase in body temperature and in respiration rate (Fig. 10A-D). Figure 11 depicts the growth curve of the Yoshida following subcutaneous implantation of 100 mg tumour homogenate into the dorsum of the foot. When the tumour was subjected to a 1 h period of hyperthermia on the exponential part of the growth curve at day 9 (maximum tumour volume approximately 1.5 ml), the tumour regressed completely and disappeared within 14-15 days. A 2 h heating period at this time led to tumour regression within 6-7 days. As the tumour increased in volume, a longer heating time was required for destruction. At a volume of 2.0-3.0 ml, a 1 h heating period at 12 days led to only a temporary restraint and decrease in volume followed by return to an exponential rate of growth (Fig. 11). The tumour could be cured by a 2nd hour at 42°C applied up to day 16 before rapid increase in volume occured again, or by a 2 h heating session on day 12.

These results show that the Yoshida tumour can be destroyed by 42°C in vivo as predicted by in vitro assay. For tumour destruction, heating must be adequate in terms of tumour mass, i.e. the response is dose-dependent. The response of the Yoshida tumour to inadequate hyperthermia (Fig. 11) is similar to the dose-response curves of solid tumours to inadequate radiotherapy or chemotherapy. After the initial regression period, 'recurrence' takes place and tumour volume again increases at about the same rate as before treatment. The complexity of this recovery process in terms of cell death, lysis and kinetics of surviving cells in relation to the tumour growth curve have been discussed in detail by SKIPPER (160). If the heating is inadequate, not only does the Yoshida

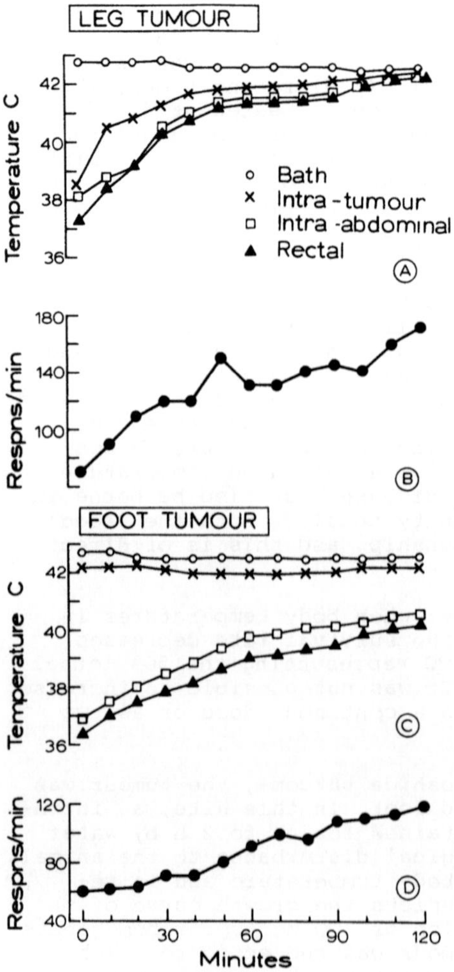

Fig. 10A-D. Temperature and respiration profiles during local hyperthermia for rat with Yoshida tumour (volume 10.3 ml) implanted in hind leg muscles (A and B) and for a rat with tumour (volume 1.5 ml) implanted on the dorsum of foot (C and D)

tumour resume exponential growth, but as illustrated for the neoplasm in the leg, the cells become altered in a way that leads to enhanced dissemination of the cancer. In previous in vitro-in vivo hyperthermia studies on SDB cells, it was found that cells not killed by exposure to 42°C recovered rapidly and completely (39). Recently, PALZER and HEIDELBERGER have postulated from in vitro experiments on HeLa cells that hyperthermic killing is at least a two-step process; for cell destruction, sublethal injury must be converted into lethal damage, otherwise the cells recover (131). The importance of reversibility in relation to the deleterious effects of heat on cancer cell metabolism has already been discussed.

The pattern of response to elevated temperature displayed by the animal tumours used in the present research has been described as that shown in Figures 1 und 8. The charts indicate that inhibition of metabolism occurs at 41.5°C and above; the tumours in the host regress when treated at 42°C. At temperatures below 41.5°C but above normal body temperature, there is a heating range over which metabolism is increased; this stimulation is most marked at 40°C. Since the pattern of these temperature charts (Figs. 1 and 8) forms the basis of our

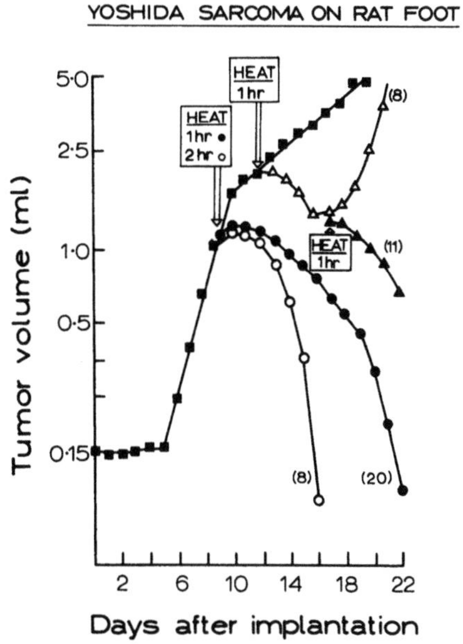

Fig. 11. Relationship between tumour volume and duration of heating (intra-tumour temperature 42°C) required to produce tumour regression

approach to the treatment of tumours in the animal, it was necessary to determine the significance of the stimulatory temperature range in terms of in vivo heating. To this end, we maintained Yoshida tumours on the feet of rats at 40°C for 1 h; the tumours were heated on day 9 after transplantation when the volume was approximately 1.5 ml. Heating led to an increase in both respiration and glycolysis of the tumours, and this was followed by enhanced dissemination of the cancer (37). From an implantation site in the foot, the Yoshida sarcoma followed a pattern of spread that was different to that consequent to leg muscle inoculation. The foot tumour metastasized more slowly with a preference for lymphatic spread, direct spread into the abdomen and vascular metastasis to the viscera being less common. Spread of the tumour was studied at 15-16 days and 26-27 days after transplantation, i.e. 6-7 days and 17-18 days after heating at 40°C. At these time points, there was a significantly increased involvement of superficial and deep lymph nodes by tumour and the incidence of tumour in the viscera (small intestine, heart) was greater than in control unheated rats. In the heated animals, tumour was also found at sites not involved in the control series, notably the testis and stomach. Figure 12 illustrates the extensive involvement of lymph nodes and viscera, and also direct spread (from an extension in the leg muscle) into the posterior abdominal wall after the foot tumour was heated at 40°C for 1 h.

It is therefore seen that the effect of heat on the malignant cells in vitro was again mirrored in vivo. In this case the action was a stimulatory one that when applied in vivo led to augmented spread of the cancer. Over the 1 h heating period (intra-tumour temperature 40°C), the central body temperature of the rats was not elevated; on immersion of the tumour-bearing foot into the bath body temperature was usually about 36°C due to the anaesthetic, and after 60 min hyperthermia the temperature had increased to the region of normal (38°C). These experiments would appear to indicate, therefore, that the accelerated spread of the tumour was in this case occasioned by some change in the cells of the primary tumour mass at elevated temperature; if the normal host tissues were altered and became more hospitable to the cancer cells, a direct action of heat as the mediator is unlikely.

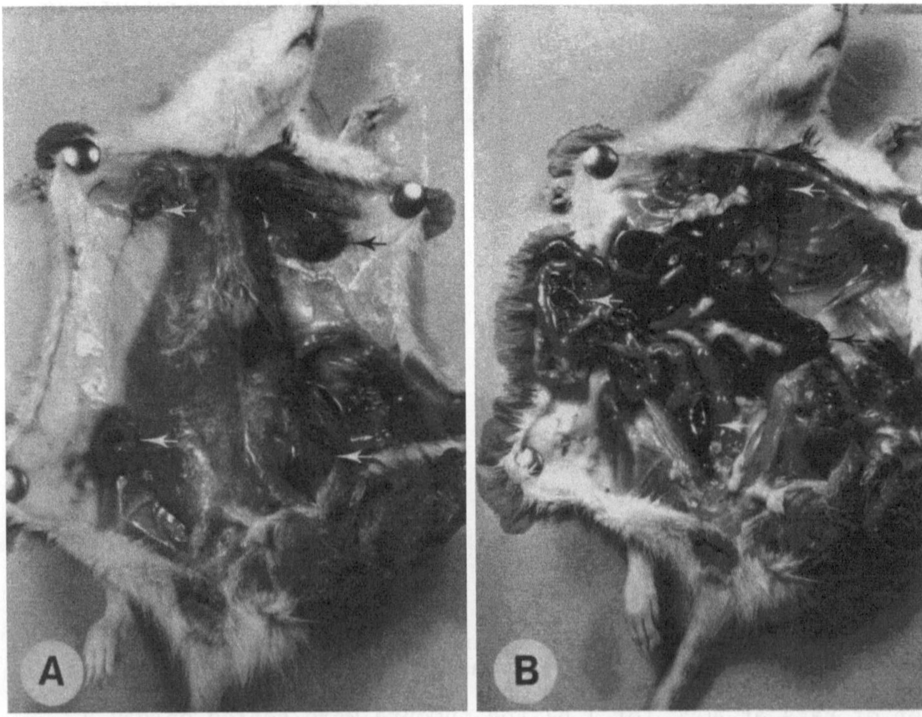

Fig. 12A-B. Gross enlargement of axillary and inguinal lymph nodes by tumour (A) in rat that died 1[illegible] days after primary Yoshida tumour on foot was heated at 40°C for 1 h. Tumour (arrows) has extended into retroperitoneal tissues and up posterior abdominal wall to level of diaphragm, surrounding kidney en route (B). There is widespread involvement of small intestine and thymus is replaced by tumour

On the basis of the data discussed, Figure 13 is presented as a working guide to the treatment of tumours in vivo by hyperthermia. Although the destructive effects of heat become more apparent as the intratumour temperature increases above 41.5°C, at temperatures over 42°C the differential susceptibility between malignant cells and normal cells decreases and host tolerance becomes a prime consideration. Temperatures below 41.5°C have been reported in the older literature to be curative for tumours. Such temperatures have usually been encountered in man in the course of concomitant pyrexial illness, e.g. erysipelas in cancer patients, or as the result of toxin therapy; in such cases, it has not been feasible to separate the possible benefit of the heat from the effects of the toxin component. Heat by itself at levels below 41.5°C, even if effective as an antitumour agent, entails very prolonged exposure periods (see Figure 18). Of much greater importance is the possibility of a stimulatory temperature range before the lethal temperature range is reached; the potential hazards in thermotherapy of such a range are illustrated by the results obtained with the Yoshida sarcoma in rats. There are a few indications in the literature that the growth rate of malignant cells increases with temperature before the inhibitory temperature is reached. SELAWRY et al. (153) reported an increase in growth rate of cancer cell lines in culture up to 39°C, and RAO and ENGELBERG (138) described a decrease in the mean generation

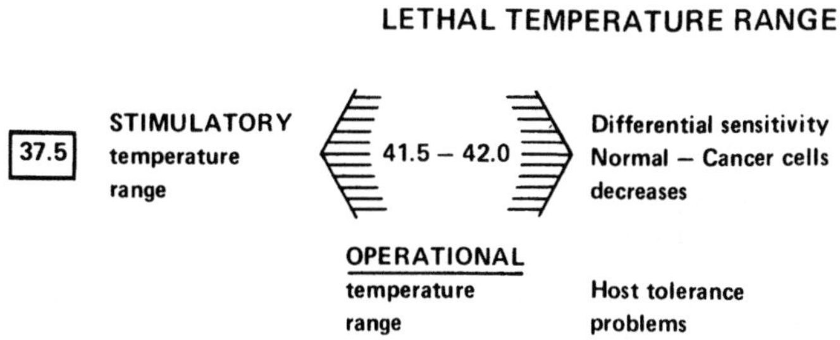

Fig. 13. Operational temperature range for hyperthermic treatment of cancer by currently available methodology in relation to stimulatory temperature zone and lethal temperature range for cancer cells

time of HeLa cells at temperatures above 37°C. The operational temperature range for cancer therapy in animals and in man is therefore a fairly restricted one, 41.5-42°C. As discussed previously, within this range the small increase of 0.5°C from 41.5°C to 42°C leads to a significant decrease in the thermal death times of tumours (see also Figure 18), 42°C being markedly more damaging to the malignant cells. In view of these various considerations, an intra-tumour temperature of 42°C would appear to be the optimal operational temperature for thermotherapy with currently available methods of applying heat to tumours.

B. Potentiators of Hyperthermia

From time to time, various substances and modalities have been advocated as potentiators of the lethal effects of heat on cancer cells. These have ranged from serotonin to radiotherapy. Since the basic requirements for optimal response to hyperthermia (e.g., exact temperature to use, duration of heating) had not been defined, it is not surprising that combination therapy has not found wide appeal.

CRILE reported that if serotonin was injected into S-180 tumours on the feet of mice before heating, then the tumours could be destroyed by half the exposure to heat that was normally required (29). This observation is an intriguing one, since an inflammatory response has been implicated in the host reaction to toxins (see below), and serotonin is a vasoactive amine that may act as a permeability factor in inflammation (177). Various less physiological substances at non-toxic doses, including methylene blue, dimethylsulphate and tween 80, have been stated by VON ARDENNE et al. to have suitable properties as preconditioners for hyperthermia (4). For his multiphase approach to cancer therapy (Krebs-Mehrschritt-Therapie), VON ARDENNE maintains that preconditioners enable the effective heating temperature for tumour destruction to be reduced to the region of 40°C, and proposes lowered pH as the best preconditioner. At a tumour pH in the region of pH 6.5, achieved by encouraging increased glycolysis in the cancer cells by maintaining an increased blood glucose, it is claimed that at 40°C tumour destruction is brought about by a chain reaction mediated by the lysosomes (4, 5). However, the published results of such approaches to cancer therapy in animals and in man are not impressive. Other workers have found that tolerance to hyperthermia is markedly lowered by the use of preconditioners (see ref. 175), and the beneficial effects of an elevated blood glucose level as an adjuvant in the treatment of cancer by total body hyperthermia in humans have been unconvincing (R.T. PETTIGREW, personal communication).

Most attention has focussed on cytotoxic drugs and radiotherapy as possible agents for combined use with hyperthermia. It has been known for over 60 years that elevated temperature can sensitize tumours to the effects of irradiation, and since the work of MULLER in the early 1900s (124), heat has been applied sporadically as an adjuvant in the radiation therapy of cancer, both in animals and in man. SELAWRY et al. (152) have reviewed results in this field up to 1958. It seems clear from this work that the effectiveness of radiotherapy may be increased by 40-60% when combined with hyperthermia as judged by the cure rate in experimental tumours. Growth inhibition and tumour regression have been produced by a combination of temperatures and radiation dosages that were by themselves ineffective, and potential application of these findings in relation to radioresistant tumours has been emphasized (152). It is also apparent that the radiosensitivity of normal tissues is also augmented by heat, but no author has reported permanent damage to the healthy tissue surrounding the heated lesion. In more recent years, CRILE (30) has confirmed the synergistic effects of heat and irradiation on several experimental tumours in mice. This worker also showed that the radioresistant S91 melanoma and T241 sarcoma became radiosensitive at elevated temperature, and that the sensitizing effects of the heat were less pronounced on the normal tissues in which the tumours grew. Results of a similar nature were obtained by OVERGAARD and OVERGAARD with a radioresistant mammary tumour in mice (128). Radiation alone (1600 rads) had no effect on the tumours while 1600 rads followed by 42.5°C for 60 min resulted in a 77% cure rate.

A beneficial synergistic action of increased temperature and cytotoxic drugs on transplantable animal tumours and also in some cases of clinical cancer has been claimed by several workers (see ref. 171 for a recent review). In view of the sensitizing porperties of X-rays, it is not surprising that the majority of these reports have concerned alkylating agents, drugs recognized to have radiometic effects. The most conclusive information on the increased effects of drugs and hyperthermia has been obtained from experiments in which tumour cells have been treated *in vitro*, and the subsequent ability to induce tumours on inoculation into animals used as an index of the value of combined therapy. By this technique, GIOVANELLA et al. (59) observed that hyperthermia in combination with L-erythro-α, β-dihydroxybutyraldehyde (DHBA) was 100-fold more effective against L1210 mouse leukemia cells than either treatment applied separately. This approach is of very limited value, since the response of the tumour in the host to heat and to drugs may vary from that in vitro (see refs. 41, 139) and no indication is obtained regarding the effect of the combined therapy on normal tissues. When used in the tumour-bearing host, it has been found that the enhanced action of cytotoxic drugs at elevated temperature is not confined to tumour tissue. SHINGLETON et al. (156) detected a significantly increased uptake of ^{14}C-labelled nitrogen mustard in normal tissues and in VX2 tumours in hyperthermic hind limbs of dogs. With regional perfusion of the hind limb in dogs, SUZUKI (171) reported that even with reduced doses of Nitromin (nitrogen mustard N-oxide) in the perfusate, irreversible damage to the limb occured at a tissue temperature of 42°C; this temperature by itself did not damage the tissues. A need for caution is therefore indicated regarding the possible action of cytotoxic drugs at raised temperature, even in the case of tumours found to be sensitive to the drug at body temperature. At hyperthermic temperatures many factors governing the activity of a cytotoxic drug may be altered, including dissociation into the active radical (e.g. the ethyleneimmonium ion in the case of alkylating agents), biological half-life, passage across cell membrances, binding to plasma and cell constituents.

In the tumour-bearing host, SUZUKI (171) reported results for treatment of Yoshida sarcomas implanted on the feet of mice. Neither nitrogen mustard nor heat individually had any influence on the tumour, but combination therapy led to tumour regression with complete disappearance in some cases. In patients, STEHLIN has reported a significantly increased survival rate for melanoma of the limb treated by regional perfusion at 40°C with phenylalanine mustard in the perfusion fluid; however, control studies using hyperthermia alone were not performed (165). Other workers have used alkylating agents at elevated temperature in the treatment of head and neck cancer (188) and of intra-abdominal tumours (157) by perfusion techniques; cytotoxic drugs have also been given during total body hyperthermia to a small number of patients with disseminated malignant disease (135). The results of these trials are difficult to evaluate, but a convincing increase in patient survival due to the combined treatment is not apparent. Although the idea of a synergistic anti-tumour effect of cytotoxic drugs and hyperthermia has been propagated (165, 171, 175), there is little evidence to date to encourage the belief that by such combination the therapeutic ratio of many cytotoxic drugs will be usefully increased, or that agents ineffective at body temperature may be rendered active at 42°C. The mechanism of action of X-rays on cancer cells is more uniform than that of cytotoxic drugs and being a physical entity, X-rays produce a less variable and unpredictable response than anti-cancer drugs which have proven to be of much less value to date in the treatment of neoplasia. Practical appplication of combined radiotherapy and hyperthermia depends upon knowledge of optimal conditions of combination. This implies adequate control of tumour temperature as well as irradiation dose. Most reported studies have been lacking in this respect, and the clinical date have usually been inadequate and lacking controls. Nevertheless, it does appear from the literature that well-designed and executed experiments and clinical trials on the complementary effects of radiotherapy and hyperthermia have more to offer in formulating cancer therapy than similar work with drugs and heat.

In the course of the present work, the response of the VX2 carcinoma in the rabbit to combination therapy was investigated. We examined the adjuvant effect of a non-curative (sub-optimal) application of local hyperthermia in combination with a sub-optimal dose of radiotherapy or chemotherapy (methotrexate). For hyperthermia, an intra-tumour temperature of 42°C was mainted for 1 h only; radiotherapy comprised 1000 rads TD applied by superficial radiation from two opposing fields; chemotherapy consisted of methotrexate given by 6 daily i.v. injections of 0.4 mg/kg. None of these individual treatments cured the VX2 tumour. The results on the primary tumour and animal survival, when compared with those from curative local heating and curative radiotherapy, illustrate the importance of the host-tumour relationship in therapy.

Figure 14 shows the response of the primary VX2 tumour to the various treatments employed. Each method of therapy, when applied individually, resulted in a restraint of tumour growth compared with untreated tumours; the difference in volume of the control and treated tumours was statistically significant ($P < 0.001$) at the 8th week after tumour inoculation for all three methods of therapy. Subsequently, the tumours treated by heating or by methotrexate increased in volume, while the irradiated tumours decreased in volume. Following combination therapy, tumour volume continued to increase for a further 2 weeks. Thereafter, the tumours given radiotherapy plus hyperthermia, or radiotherapy plus methotrexate decreased in volume and the tumour volume measurements for both types of dual therapy were comparable at each time point ($P > 0.05$). The volumes for these tumours were significantly less than the corresponding volumes for tumours treated by radiotherapy alone

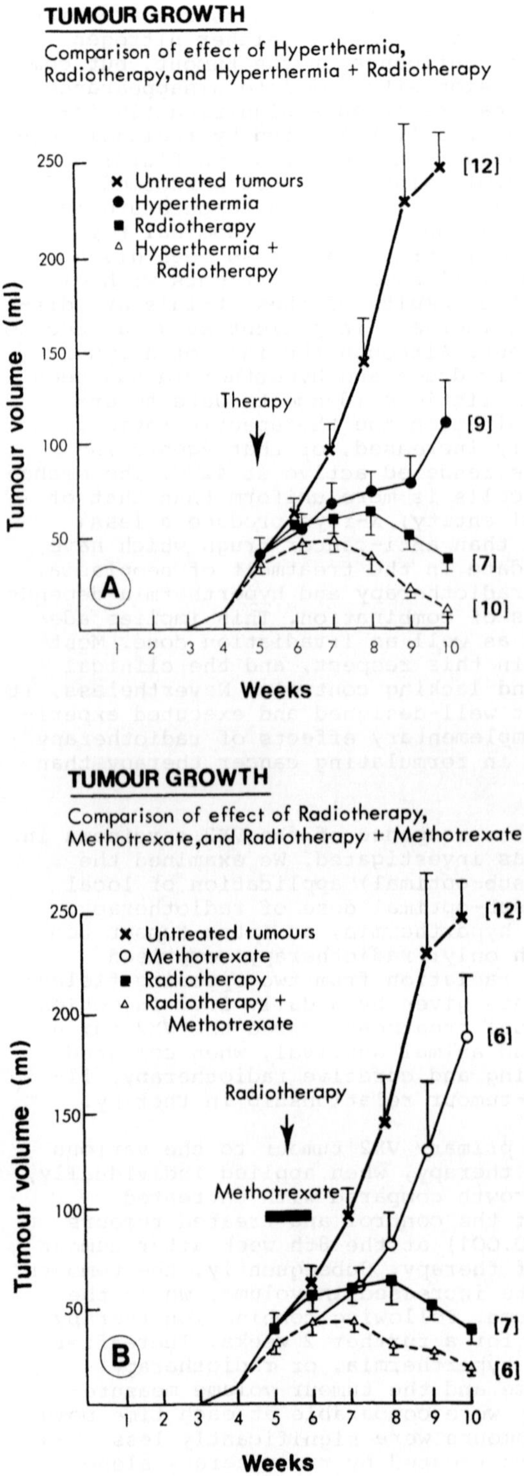

Fig. 14. Changes in mean VX2 tumour volume following combination therapy. Hyperthermia, applied by water bath immersion on day 35 after tumour inoculation, was followed within 2 h by 1000 rads irradiation to tumour area (A). For methotrexate combination (B), drug (0.4 mg/kg body wt/day, i.v.) was given on days 35-40 after inoculation, and a single treatment of hyperthermia given on day 38

($P < 0.05$). For tumours treated by methotrexate in addition to hyperthermia, the changes in volume were not significantly different from the changes in volume following hyperthermia alone (Fig. 13A).

Histological Changes. The microscopic changes following inadequate or curative heating of the VX2 tumour were described earlier. Sub-optimal radiotherapy (1000 rads) to the tumour produced central necrosis, leaving a rim of intact cells in which mitotic figures were present 4 weeks after therapy, when the necrotic central area had been replaced by fibrous tissue (Fig. 15). Increased dosage of radiotherapy (2000 rads) led to widespread tumour cell destruction and replacement by fibrous tissue. No microscopic changes were observed in the tumour following methotrexate. The potentiating effect of heat or methotrexate produced no striking alteration in the histological picture following irradiation.

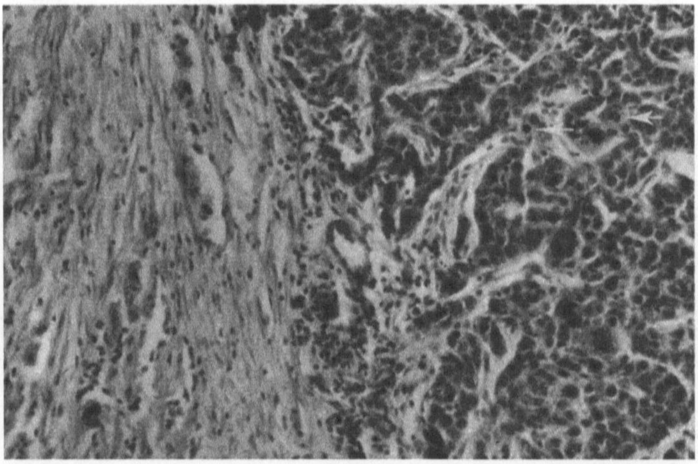

Fig. 15. VX2 carcinoma 4 weeks after radiotherapy (1000 rads), showing peripheral intact cells with mitotic figures (arrows), and central zone of necrosis replaced by fibrous tissue. x 200

Tumour Metabolism. Table 2 records the respiration of the treated tumours at 24 hours, 10 days and 4 weeks following therapy; respiration is the parameter that correlates with growth in the case of the VX2 tumour (38, 122, 123). Following each of the three individual therapies (42°C for 1 h, 1000 rads radiotherapy, methotrexate) O_2 consumption was depressed only temporarily and had recovered to control values at 4 weeks after treatment. The effects of combination therapy reflected a summation of individual therapies; there was synergism of action in the case of radiotherapy and hyperthermia or methotrexate, as indicated by persistent marked inhibition of O_2 uptake at 4 weeks. After radiotherapy (2000 rads) as after 42°C for 1 h on three occasions, the failure of respiration to recover to normal values correlated with the widespread tumour cell destruction following treatment.

It is therefore seen that we have treated the VX2 tumour in various ways, and achieved a similar inhibition of metabolism with single therapy (heat 42°C for 1 h x 3, RT 2000 rads) and with combination of synergistic therapies (RT 1000 rads + heat 42°C for 1 h; RT 1000 rads +

Table 2. Oxygen uptake (QO_2) of VX2 cells after <u>in vivo</u> therapy

Therapy	24 h	10 days	4 weeks
Untreated VX2	8.5	8.4	8.1
42°C for 1 h	3.0	5.1	7.3
Radiotherapy (RT) 1000 rads	4.4	7.7	7.5
Methotrexate (MTX)	10.2	4.7	7.6
42°C for 1 h + RT	3.7	4.7	4.4
42°C for 1 h + MTX	3.4	3.6	7.8
RT + MTX	5.1	5.0	5.5
42°C for 1 h x 3	3.5	5.2	2.8
RT, 2000 rads	3.0	4.0	3.6

Each figure is the average value from 3 separate tumours. At 10 days and 4 weeks after 1000 rads RT, the cells were obtained from the intact peripheral rim of the tumours. The 4 week values following 42°C for 1 h x 3 and RT, 2000 rads, represent the QO_2 for slices of fibrous tissue that replaced the tumours.

methothrexate). Biologically also, the response of the tumour to these four treatments was similar -- the regression lines for tumour volume from the 7th week onward after tumour transplantation did not differ significantly in slope.

When the results of these different treatments were evaluated in terms of animal survival, the composite picture shown in Figure 16 was obtained. At the time of death, all animals had metastases in the lungs, inguinal and para-aortic lymph nodes. It can be seen that, with the exception of fractionated local or total body heating, none of the therapies produced a worthwhile increase in survival time compared with the control group of rabbits that died within 70 ± 6 days. Although radiotherapy potentiated the action of heat (and of methotrexate) as evidenced from the regression rate and inhibition of metabolism in the primary tumours, the situation is not a straightforward one, in that potentiation was not equated with an advantage to the host. Adequate radiotherapy (2000 rads) destroyed the primary tumour, but the rabbits still died from metastases. Only three applications of local hyperthermia led to a significant cure rate in rabbits bearing the VX2 carcinoma.

6. DISCUSSION

A. Relationship Between Primary Tumour and Its Metastases and Host-Tumour Relationship

The series of findings just described focuses attention on the relationship between a primary tumour and its metastases.

A primary tumour controls the establishment of its metastases and once metastasis has occured, the primary exerts an influence on the number and rate of growth of its satellites. There is evidence that the response of the host and the total host burden of cancer cells are im-

Duration survival in each form of treatment

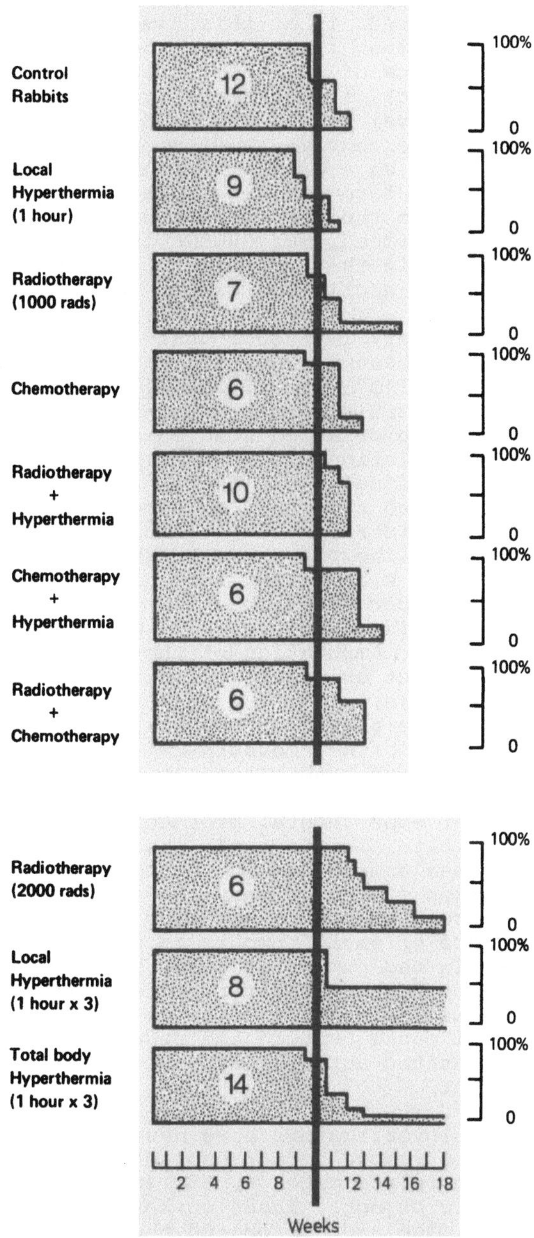

Fig. 16. Survival time of rabbits bearing VX2 carcinoma following different therapy schedules. For upper histograms, individual therapies were sub-optimal for destruction of primary VX2; for lower 3 histograms, curative treatment was applied to primary tumour. Solid black line indicates mean survival time in untreated rabbits (70 ± 6 days), and figures in circles show numbers of animals in each group

portant in these relationships. CRILE (32) has shown that in inbred mice with a metastasizing tumour implanted on the foot, if the foot is amputated before day 12, metastases rarely appear and most of the mice are cured. If at this time blood is taken from the aorta or vena cava or if lung tissue from the mice is injected into the thigh muscles of other mice of the same strain, the tumour is nearly always transferred. Hence, though cancer cells may be present in both blood and lungs, removal of the tumour may still prevent metastasis and effect a cure. From extensive studies on transplantation and metastasis, GREENE showed that as a primary tumour evolves, it brings about a conditioning of the host favouring its spread (67, 70). The constitutional changes concerned in the differing behaviour of the tumour-bearing animal have not been defined, but GREENE and HARVEY subsequently postulated an alteration in the receptivity of the vascular endothelium to tumour cells to be a critical factor (68, 69).

The inhibitory influence that a primary tumour may exert on its metastases has been known since the beginning of the century when MARIE and CLUNET (107) and TYZZER (173) reported enhanced growth of metastases following partial excision of implanted tumours in mice. Subsequently, several more detailed studies have supported the concept of a controlling relationship between primary and secondary tumours. SCHATTEN (149, 150) studied the pattern of metastasis of S91 melanoma and DBA sarcoma 49 to the lungs of mice after surgical removal of the limb bearing an intramuscular tumour implant; latent metastases were present at the time the primary was removed. At 21 days following amputation, there was a significant increase in the mean number and mean size of metastases per mouse obtained with both tumours, and with the sarcoma there was also an increase in the number of animals (frequency) showing secondaries. Amputation of the leg opposite the tumour-bearing limb had no effect on growth of metastases. SCHATTEN concluded that with these cancers, after the primary tumour attained sufficient size, it retarded the growth of established metastases and also inhibited the development of new metastases. KETCHAM et al. (93, 94) studied metastasis in five transplantable tumours in mice, including S91 melanoma, sarcoma 49 and sarcoma 241. Using intramuscular tumour implantation and a similar experimental protocol to SCHATTEN, these workers were unable to duplicate his findings. A significant decrease in the frequency of metastases and in the number of metastases per mouse occurred following amputation of the primary. However, although removal of the primary mass brought about a cessation of tumour cell shedding and reduced the incidence of small tumours in the lungs, there was a consistent increase in the number of large pulmonary metastases. This phenomenon was interpreted as an indirect effect reflecting the better condition of the animals with amputation or the lesser competition for nutrient material consequent to the reduced number of metastases, but stimulation of metastatic activity by removal of the primary tumour could not be ruled out.

Several investigators have described an increased incidence of metastases following removal of transplantable tumours from subcutaneous sites in hamsters and mice (6, 68, 101) and the experiments of LEWIS and COLE (101) are of note. These workers used the T241 sarcoma implanted on the feet of mice and corrobated the findings of SCHATTEN that as a primary cancer increased in size, it exerted a restraining influence on the occurence of metastases. If the primary T241 tumour was removed on the 10th day after inoculation only 7% of the mice had pulmonary metastases at death, whereas 36% of the controls had metastases. When the primary tumour was left in situ until the 14th day or later, then its removal resulted in a significant increase in the frequency of secondaries and also in the number of these per host (e.g. at the 22nd day the incidence of metastases was 57% in control animals versus 96%

in the amputated series). Assuming that cells were disseminated to the lung in equal incidence in the two series, and that the primary tumour remained capable of shedding cells until the death of the animal, it was difficult to escape the conclusion that the metastatic cells were killed by the tumour-bearing host (101).

Increased incidence of metastasis has also been found following the removal of spontaneous tumours. The reports have concerned the spontaneous mammary tumours of mice (13, 74, 103), and the work of LIEBELT et al. (103) is of special interest. After surgical removal of the primary tumour, 80% of the animals had pulmonary metastases 60 days after operation compared to an incidence of 39% in the control series. In addition, an increased number of tumours arose in the mammary glands after removal of the original primary tumour.

Examples illustrating that a primary tumour may influence the development and growth of its secondaries are well-recognized clinically. FERGUSON (47) noted that the time of appearance of pulmonary metastases from osteogenic sarcomas of the extremities correlated more closely with the time of amputation than with the time of first appearance of the tumour, even when there was considerable delay in treatment. It must be remembered that the influence of the primary may be stimulatory to metastases. For example, regression of secondary tumours following removal of primary human cancers of certain organs such as the breast and kidney is well-known to therapists.

CRILE (32) has indicated the potential role of the host immune system in the relationship of a primary tumour and its metastases from experiments with an allogeneic tumour in mice. When S180 was implanted on one hind foot of mice, the animals soon became immune to reimplantation of the tumour. However, if the tumour-bearing foot was amputated on day 10, and 4 days later the contralateral foot was challenged by a dose of 80 x 10^4 tumour cells, 60% of the mice were susceptible to reimplantation. Hence immunity to reimplantation of the tumour disappeared rapidly after the primary tumour was removed. On the other hand, if the S180 was treated by a single destructive dose of irradiation, immunity to the tumour persisted for at least 3 weeks. With another mouse tumour system, the incidence of metastasis was almost doubled when the foot bearing the tumour was amputated compared to when it was irradiated. From these data, CRILE suggested that the cells killed by irradiation maintained antigenicity and prolonged the animal's immunity to the growth of secondaries (32). In further work, CRILE supplied laboratory and clinical evidence to support the view that, when there is an immunological reaction of the host against a cancer, the primary tumour may act as the antigenic source necessary to maintain immunity against metastases (31). With both syngeneic (sarcoma 1) and allogeneic (S180) tumours in mice, it was found that removal of the regional lymph nodes increased both the tendency to metastasize and the implantability of the tumours compared with removal of the tumour only. Similarly, there was suggestive evidence that immunized regional nodes played a role in preventing the dissemination of human breast cancer, and that such nodes should not be removed at mastectomy unless they were involved by tumour (31). MITCHISON (116) had previously demonstrated that early in the growth of a tumour in mice, tumour immunity could be transferred by transplantation of lymph nodes draining the tumour.

The balance between primary and secondary cells in the host may be a finely adjusted one according to the analogous work of VANWIJCK et al. (174). Immunogenic sarcoma cells were inoculated into the rear limb muscles of syngeneic mice, and following amputation of the limb at 20 days (tumour volume 2-4ml), 50% of the mice died from pulmonary metastases at 81+ days. Complete protection from metastases was provided by injecting 1 x 10^4 live, syngeneic tumour cells into the intact contralateral leg at the time of amputation. When the dose was increased to 5 x 10^4 cells, all the animals died of metastases within 43 days. Irradiated cells at either of these doses also stimulated the development of metastases. Alteration of the cells by irradiation or a small increase in tumour cell inoculum was therefore sufficient to change completely the host response to the cells already shed from the primary tumour. VANWIJCK et al. believed that in these experiments the cause of the increased tumour growth was the additional antigenic load on the host (174).

There is therefore considerable evidence that a primary tumour influences the occurence and growth of its metastases. Increasing attention is being directed to the view that the development and progress of a given neoplasm is influenced by the relationship operating between the growing tumour and the reaction of the host (52, 67, 103, 145). Although the nature of the host factors has not been precisely defined, it is important to emphasize that such factors are constitutional in distribution and not localized at the site of the primary growth (67, 103). The fundamental role of the immune system in this context is becoming more apparent. In the therapy of cancer, we are disturbing this complex inter-relationship between the host, the primary tumour and its metastases, and it is essential to consider the effect of treatment not only upon these components but also upon their inter-relationships. The manner and extent in which these relationships are perturbed, and the subsequent effect upon the host, will depend upon the precise mode of therapy (surgery, drugs, immunotherapy, heat) and the degree to which it affects primary and metastatic cells as well as the host. The response to treatment is therefore governed by host-tumour-therapy considerations. The importance of the host element in determining the therapeutic outcome of the host-tumour-therapy triad when cytotoxic drugs comprise the therapy is well accepted by chemotherapists (64, 92). At the present time, knowledge of the inter-relationships of this triad, and how best to control them for the benefit of the host, is only too meagre. The host-tumour relationship is not a static one but changes as a primary tumour evolves and progresses to disseminated disease; concomitantly the disease becomes a more difficult therapeutic challenge. The reduction in therapeutic effectiveness against disseminated disease involves alterations in the tumour component of this triad (e.g. increase in the total number of malignant cells, changes in the kinetics of the tumour cell population, development of resistance to drugs, different metabolic characteristics of metastatic cells) and also modifications of the host component (e.g. toxic manifestations of the cancer on normal cells, decreased capacity of the immune system) (65).

The present results with the VX2 carcinoma in the rabbit are interpreted in terms of the tumour-host relationships just discussed. The rabbit was treated at a time when the disease was widespread with metastases in the regional nodes, distant nodes and in the lungs. Destruction of the primary tumour by radiotherapy did not significantly affect survival time of the animals. Adequate local hyperthermia led to total dissolution of the primary tumour with eradication of metaseses (Fig. 17). It was postulated that this occurred via an immune response that may have been invoked by the specific method of tumour cell death and resorption resulting from hyperthermia. Combination heat and radiotherapy destroyed part of the tumour mass. Although this may have led to stimulation of the immune system, in terms of antigenic load the response may have been inadequate to cope with the total remaining host burden of cancer cells. Evidence for the importance of a minimal host burden of tumour cells for an effective immune response has been provided by the work of MATHE on the immunotherapy of leukaemia (109).

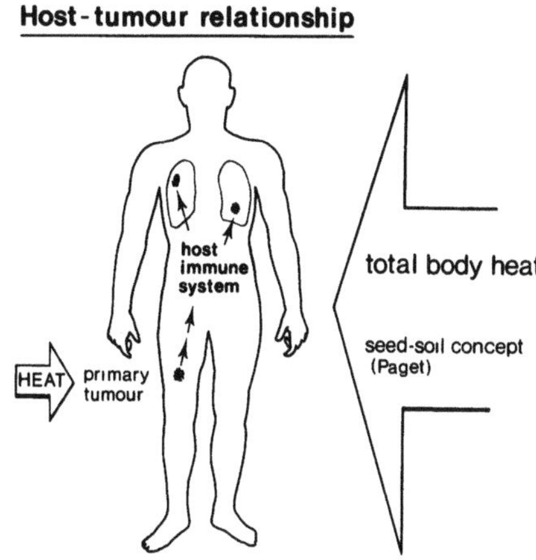

Fig. 17. Local hyperthermia and total body hyperthermia viewed in terms of effects upon primary tumour, metastases, host immune system and other normal tissues ('soil') of host

In the total body hyperthermia, the elevation of temperature in the normal tissues and organs ('soil') of the host adds an extra reactant to the host-tumour system (Fig. 17). Efficient total body heating would be expected to destroy all malignant cells, if these are heat sensitive. The problem of different heat sensitivity which may exist between primary and secondary tumour cells should be surmountable by prolonged duration of hyperthermia. On the other hand, there are hazards involved in the inadequate heating of tumour cells as discussed earlier. In addition, after total body hyperthermia, surviving cancer cells would recover in an environment rendered more favourable to their replication and growth due to the increased metabolic rate of the heated normal tissues. The immune system may prove unable to cope with such hyperactive tumour cells; there is also the suggestion that at elevated temperature the immune response may be depressed.

In 1954, GREENSTEIN (71) stated: "the host-tumour relationship is the key to the cancer problem". Today, 20 years later, the complex nature of this relationship is becoming more apparent and its importance in therapy is receiving increasing attention. Local hyperthermia and total body hyperthermia disturb the relationship with different emphasis on the various components involved in the interaction (Fig. 17), the main distinguishing feature being the involvement of all the normal tissues of the 'soil' in systemic heating. This engenders different problems from the point of view of effective tumour cell destruction. These problems are such that local and total body heating should be regarded as different approaches to the treatment of malignant disease. It is clear, however, that hyperthermia (local and total body) represents a two-edged weapon that may alter the host-tumour relationship to the benefit or to the detriment of the host.

B. Bacteria, Hyperpyrexia and the Immune System

Participation of the immune system in the response of tumours to heat was mooted by earlier workers in the field. COLEY, who devoted a lifetime to studying the response of human cancer to induced pyrexia, found that some patients became immune to injected bacterial toxins and inevitably the possibility arose that the body's defence mechanisms were operating against the tumour as well as against the injected toxins (125). In reviewing the heyday of the application of heat to malignant tumours in animals and in man (1880-1920), WESTERMARK in

1927 (183) stated that, in relation to the treatment of tumours by heat "... one must also consider the possiblity of the organism itself being able, through its own healing capacity, to deal with tumourous tissue, damaged by heat". From the precise reference to tissue 'damaged' by heat, and the fact the JENNER's classic studies on vaccinia had been known for over 100 years, that METCHNIKOFF's book on the reticulo-endothelial system (now regarded as a major component of the immune system (191)) was published in 1890, and that BCG had already been in use for 2 years as an immunizing agent, it is difficult not to interpret WESTERMARK's comments as having an immunological connotation. There is now more definitive, albeit circumstantial, evidence implicating the immune system in the response of tumours to heat. Following destruction of tumours in the rat and rabbit by local heating, metastases disappear and the animals live a normal lifespan (38, 41). This is paralleled by the report of GOLDENBERG and LANGNER (63) that, in animals bearing bilateral tumours, there is impaired growth of tumour after heat treatment of the contralateral tumour (the so-called abscopal anti-tumour action). Analagous clinical findings are evident. In the series of 22 advanced melanomas and sarcomas (tumours with a high tendency to metastasize) subjected to hyperthermic regional perfusion by CAVALIERE et al. (19), only one patient died of metastases. In a few cases of multiple localization of tumours, heat treatment to the limb tumour resulted in disappearance of the others also. For soft tissue sarcomas of the limb, STEHLIN (166) has greatly reduced the necessity for amputation by including hyperthermic perfusion in his treatment regimen of perfusion and irradiation. STEHLIN has also been impressed by the radiological disappearance of lung metastases following treatment of primary limb tumours by hyperthermia (J.S. STEHLIN, personal communication). With recurrent melanoma and sarcoma of the limbs, CAVALIERE currently obtains best results by delaying limb amputation until 1 month after hyperthermic perfusion (see Chap. 5), the inference being that during this delay period, tumour breakdown products provoke a beneficial immune response against the residual cancer cells in the host. Although STEHLIN found a synergistic action of anticancer drugs and elevated temperature (165), in CAVALIERE's experience (Chap. 5) the addition of such agents (usually alkylating agents) to the perfusion fluid resulted in tumour recurrences more frequently than when hyperthermic perfusion alone was employed. This may be ascribed to the potent immuno-suppressive effects of alkylating agents (72); in CAVALIERE's cases the longer perfusion times (up to 8 h compared to 2 h in STEHLIN's work) may have enabled an adequate concentration of drug to leak into the general circulation to modulate the immune response. Recently, MONDOVI et al. (118) found that immunization of Swiss mice with Ehrlich ascites cells was more effective when the cells had been exposed in vitro to 42.5°C for 3 h. It was concluded that heat treatment led to an increased immunogenicity of these cells, and that this may have relevance to the mechanism of tumour regression following hyperthermia. Using synchronized monolayer cultures of a malignant line derived from rat breast adenocarcinoma cells, we have demonstrated increased binding of cell specific antiserum to the cell membrane following heating in vitro for 1 h at 42°C (88). The enhanced binding occurred during the G1 phase of the cell cycle, when the expression of cell surface antigens is maximal in relation to the other cycle phases (23).

Total body hyperthermia, in the form of hyperpyrexia, is the approach that has produced most impressive results for advanced malignancy in the past. The work of the American physician W.B. COLEY is most recognised in this respect. From 1891 to 1936 COLEY induced hyperpyrexia in cancer patients with a variety of preparations ranging from live erysipelas bacteria to extracts and filtrates of bacterial cultures. COLEY found that there was synergism of effect between an extract of the haemolytic streptococcus S. pyogenes and an extract of Bacillus prodigiosus (Serratia marcescens), and the mixed filtrates of these bacilli became known as Coley's toxins or Coley's fluid (now commonly referred to as mixed bacterial toxins, MBT). A critical and detailed case analysis of 30 inoperable and histologically confirmed tumours that regressed completely following therapy by Coley's toxins has been published (125); there was recurrence of the cancer in only one of the patients, some of whom were examined at follow-up for almost 50 years. In COLEY's lifetime, many physicians treated inoperable cancer by MBT inoculation. Although striking cures were recorded for different types of cancer, the toxins were rarely, if ever, standardized biologically and consequently the effects were capricious and inconsistent. In addition, several different commercial preparations (e.g. the Buxton and the Tracy preparations) were in use. In more recent years, a number of investigators have studied many different microbial products that might prove effective in the treatment of malignant disease. It is of note that although the organisms examined range from Escherichia coli to B. pertussis (see 114 for refs.), no preparation has emerged to rival the MBT of COLEY.

The older literature indicates that although cures were recorded following cancer treatment with toxins, in the majority of patients tumour regression was incomplete and the remissions were of limited duration. It was not until the end of his life that COLEY realized the great importance of standardizing not only the MBT but also the technique of administration. He attributed the diverse and inconsistent results obtained by his contemporaries, and also by himself, to differences in site of inoculation of the toxins, variations in dosage schedules and duration of therapy (125). COLEY considered the most important factor for success to be adequacy of toxin therapy. Patient survival rates increased with duration of therapy and COLEY believed toxin injections should be continued long after all signs of cancer disappeared; treatment should be continued over a period of months or years rather than weeks (125). The value of prolonged toxin treatment in malignant disease has been confirmed by more recent clinical experience (114).

Successful treatment with the bacterial products led to the production of a high fever, tumour haemorrhage and tumour necrosis. In the early 1940s, SHEAR and his group (155) isolated and purified from B. prodigiosus a polysaccharide fraction that produced haemorrhage and necrosis in spontaneous and transplantable tumours, both carcinomas and sarcomas, of mice. Although the action of the polysaccharide was selective for tumour tissue, in most instances complete and permanent regression of the tumours did not occur following treatment of the mice. Repeated injection of the preparation failed to destroy remaining viable cancer cells and the tumours usually resumed growth and killed the host animals. In 1949 REIMANN and NISHIMURA (140) carried out clinical trials on about 100 patients with a wide variety of tumours, using SHEAR's polysaccharide. Although this extract is often referred to as the 'tumour-necrotizing' fraction, injection into patients resulted not only in some haemorrhage and necrosis in all tumours except melanomas but also in a fever in the region of 105°F lasting 2-6 h. The results were quite irregular. In some patients, marked reduction in tumour size occurred within 10-14 days of polysaccharide administration; in others, the material had no influence on the tumour. As in the case of mice, complete destruction of tumour was not achieved in any patient, and the rapid development of immunity to the polysaccharide precluded its repeated use. When tested clinically by other workers the purified polysaccharide proved highly toxic (139, 154). Purification of an active agent from MBT therefore turned out to be self-defeating and the results of its administration were uniformly poorer than those obtained with the crude Coley's toxins.

It is of note that the refined toxin as prepared by SHEAR retained both tumour necrotizing and pyrexial properties. BRUNS as far back as 1887 (14) recorded elevated temperature (104°F for several days) in a woman whose sarcoma of breast and its metastases regressed following an attack of erysipelas. COLEY, in his early work between 1893 and 1906 (27) noted that the body temperature rose to 104-106°F after each toxin injection and that patients with the highest fevers had the longest remissions. It was left to VIDAL, however, in 1911 (176) to conclude explicitly from a review of work up to that time that the various beneficial effects of bacterial toxins were due to the fever they produced. Much of the subsequent work in the field of hyperthermia was directed towards the elevation of tumour temperature by physical means, and in the recent revival of interest in hyperthermia, the involvement of the bacterial factor has been ignored. There is now convincing evidence that under the right conditions heat has a selective and deleterious effect on some types of cancer cells (12, 19, 29, 35, 175). What this means in terms of the individual patient with a specific cancer is largely unknown at the present time. Satisfactory techniques of general application for heating tumours have not been developed; we therefore do not know to what extent sarcomas and carcinomas will respond to adequate heating, e.g. 20 h at 42°C (35). The treatment of tumours by regional perfusion (19, 165) and by total body hyperthermia (78) as currently practised are specialized, time-consuming techniques and the results to date are inferior to the use of bacterial toxins (see 114 for refs. to the recent New York Cancer Research Institute monographs by G.A. FOWLER on the end results of toxin therapy for various types of human cancer). It may be significant that complete regression of tumours has occurred most frequently in patients with high fever due to the presence of bacteria, e.g. erysipelas (see the above monographs by FOWLER) and that COLEY achieved his most consistent results in his early series of patients who were treated by injections of live erysipelas bacteria, dead bacteria or very crude bacterial extracts (see 125). At the end of a lifetime devoted to the study of induced hyperpyrexia by bacterial extracts, COLEY himself was undecided as to the part played in tumour remission by the direct action of bacteria or their products, the response of the host immune system and the elevated temperature.

In recent years, a number of significant studies at the experimental level have confirmed that certain types of concurrent acute infection may have a markedly beneficial effect on host resistance to cancer. For example, CHRISTENSEN (22) subjected rabbits bearing the Brown-Pearce carcinoma to massive infections with haemolytic streptococci (0.5-1.0 x 10^9 organisms) and reported that the infected rabbits developed no metastases, while 50% of the control animals had metastases; the rabbits were given a protective dose of penicillin 3 days after the infection. ZWAVELING (192) concluded from his experiments on implantation metastasis in mice that tumour cells have less chance of growth in a milieu in which there is suppuration from bacterial infection. The effect of bacterial infection on tumour cell contamination of operative wounds was examined by MILLER and KETCHAM (113). Skin wounds in mice were injected with sarcoma cells followed 30 min later by inoculation of bacteria (haemolytic streptococci, haemolytic staphylococci, *E. coli* or *Proteus*) into the same site. Tumour incidence was significantly reduced by the streptococci and *E. coli* infections compared to the incidence in the control mice (56% and 68% respectively vs. 96% in controls). These workers observed that intimate association between bacteria and tumour cells appeared to be necessary for suppression of tumour growth. Infection at a distant site had no effect, thus tending to exclude systemic action or non-specific host resistance factors in this system. In patients, Coley's fluid was most effective when injected directly into the tumour mass, although responses were often obtained after intramuscular or subcutaneous administration (125). It is interesting that while modern surgeons regard intercurrent infection as undesirable in patients with malignant disease, the older surgeons were well aware of the possible beneficial effects of local suppuration or systemic infection in such cases (33, 189).

The very high potency of SHEAR's active polysaccharide in producing marked tumour haemorrhage led to the view that the agent produced its effects by a toxic action on the capillaries of the tumour (158). However, the comparatively few histological studies made on such tumours suggested strongly that the bacterial filtrates also had a direct, highly destructive effect on the malignant cells (2, 3, 75, 170). More recent support for these findings comes from the work of KOSHIMURA et al. (97), who found that the take of Ehrlich ascites carcinoma in mice was prevented by mixing the tumour cells with live group A haemolytic streptococci prior to transplantation. The mice were protected from infection by treatment with penicillin.

To produce these effects on injection into the tumour the organisms do not need to be alive, but a prerequisite would appear to be significant contact between the target tissues and the bacteria or their products. As could be surmised from experience in vaccine preparation, if live bacteria are not used then the method of killing the organisms and preparing the toxins is important. In CHRISTENSEN's experiments on the Brown-Pearce carcinoma, only lysates obtained by the action of a phage on the streptococci prevented tumour metastasis; heat-killed or formalin-killed bacteria were ineffective (22). The importance of such factors was well known to COLEY. In COLEY's lifetime, and subsequently, at leat 14 types of MBT were employed (125). The different methods of culturing the organisms, and of mixing and sterilizing the preparations had a considerable effect on their potency and toxicity (e.g. the amount of each organism present, the time and temperature of heating were critical; heat sterilized preparations were more immunogenic than those sterilized by filtration (125). Due to these factors and the now widely recognized variablity of microbial cultures (139), a consistently effective preparation has eluded the protagonists of MBT. In CHRISTENSEN's work, the bacteria or lysates were given intravenously and only the incidence of metastases was influenced; the primary tumours remained unaffected in all cases. From a series of investigations, ZAHL and HUTNER (190) concluded that tumour-haemorrhage potency was essentially a characteristic of all gram-negative bacteria. The discovery of a tumour-inhibitory substance in gram-negative bacteria led to the purification of the polysaccharide from B. prodigiosus by SHEAR (155). In spite of reports that agents active against animal tumours are present in E. coli (155), and that gonococcal vaccine injection was associated with arrest of tumour growth and some regression in 9% of 113 patients with advanced cancer (148), the extensive literature on the use of toxins appears to indicate that clinically the most marked and important antagonism is between haemolytic streptococci (gram-positive) and human tumours.

In addition to provoking an inflammatory response on inoculation into a tumour, bacteria may generate an anti-tumour response in the host immune system. There is evidence that streptococci, staphylococci and their toxins can stimulate a host response to substances or tissues that do not normally elicit such a reaction (167). Many years ago, BURKY (15) found that by combining staphylococci with lens substance (normally non-immunogenic), rabbits were sensitized to the lens and developed high precipitin titres for lens tissue.

More recently, it has been reported that pathogenic strains of streptococci and a strain of Staphylococcus aureus, when grown on agar, gave rise to agar antibodies in antisera prepared against the bacteria (61). The coccal toxins, which include haemolysins, fibrinolysin and hyaluronidase, can also be envisaged as acting to dissolve the surrounding protective matrix of the tumour cells, thus rendering them more available for destruction by the immune response.

Recently, MILLER and NICHOLSON (114) published an excellent detailed analysis of all but two (whose case histories were lost) known cases of histologically confirmed reticulum cell sarcoma of bone treated prior to 1956 by toxin therapy alone or combined with surgery and/or radiotherapy (47 cases), or in whom concurrent acute infection, inflammation or fever occurred (5 cases). The 5-year survival rate in the 47 toxin treated cases was 64%, almost twice as high as the reported rates in other series, and the 5 infection cases were free of disease from 8 1/2 to 44 years after onset. Sixteen of the 47 patients already had metastases when toxins were initiated. MILLER and NICHOLSON's analysis revealed that if toxin therapy was begun before or concomitantly with radiotherapy (to primary tumour and metastases) the response was prompt and dramatic with a high incidence of complete and permanent regression. If a considerable amount of radiation was given prior to toxins, the patient response was usually unsatisfactory. The authors believe that preliminary irradiation inactivates the reticuloendothelial and lymphoid tissues, thus decreasing the resistance of the patient to his tumour; on the other hand, preliminary toxins, given prior to or in association with radiotherapy, stimulate and protect these defensive tissues against the effects of subsequent irradiation (114). This opinion is supported by studies indicating that preliminary administration of bacterial toxins protects the host animal or patient's normal tissues against radiation damage while potentiating the response of tumour to irradiation (1, 20, 25, 43, 83, 163, 193).

The history of toxin therapy in cancer has many parallels in the current interest in bacille Calmette-Guerin (BCG). In this case, the emphasis is on the immune response, and BCG is being investigated as a stimulant of the immune system with a view to its use in the immunotherapy of neoplasms. The similarities between the features of this gram-positive organism and the characteristics of the response to the streptococcal toxins of COLEY's era are very striking:

1. The living, intact BCG microorganisms are more effective than killed organisms in eliciting a response (191). Crude uncharacterised extracts of the bacillus (purified protein derivative, PPD, and methanol extraaction residue, MER), appear from limited experience in their therapeutic use to date, to be equally or more effective than the parent agent (96, 179, 180, 191). COLEY's most consistent results were achieved from injections of live S. pyogenes, dead bacteria or crude bacterial extracts (125).

2. The deleterious effect of the BCG, like that of COLEY's fluid, is selective for malignant tissue, although a necrotizing ulcer can develop in the skin after BCG injection (120, 191). This destruction of normal tissue may be due to the cellular immune response to BCG, however, rather than to the bacillus per se (191).

3. As with Coley's toxins, the response to BCG is influenced by the strain of organism used, the amount given, the route of administration and the dosage schedule (75); a wide spectrum of tumour types responds to BCG, and best results are obtained with intralesional injection (76, 96, 191); there is often regression of secondary tumours following treatment of the primary mass (76, 96, 121, 191); complete regression of tumours in animals and in man often requires an aggressive dosage

schedule with the emphasis on repeated injections over a prolonged period (see above refs.).

4. As with streptococcal extract, combination of BCG extract (PPD) with a second agent (e.g. pertussis extract) increases the oncolytic potency of the preparation (96). This effect is also achieved by concurrent use of PPD and streptococcal extract (streptokinase-streptodornase). Such combinations induced tumour regression in a variety of human cancers, when the individual agents were ineffective (96).

5. BCG injection into patients can cause fever (120, 121). It is difficult to obtain data on the incidence and severity of this hyperpyrexia since it is regarded as an adverse effect of BCG immunotherapy, but prolonged fever has been reported (121).

The histopathology of BCG-mediated tumour regression has been studied in some detail. Essentially, an acute inflammatory reaction occurs 24 h after BCG inoculation into the tumour (76). This is followed by a chronic inflammatory response with granuloma formation, which is regarded as the sine qua non for tumour suppression (76, 191). As a result of lymphocyte sensitization and the arming of magrophages, the granuloma becomes the site for congregation of activated macrophages which are believed to be responsible for death of the tumour cells (96, 120, 191). The response to BCG is regarded as a delayed-type hypersensitivity reaction involving a non-specific immune response followed by the generation of a tumour-specific cell mediated immune reaction (76, 96, 121, 191); tumour specific antibodies may also play a part in the reaction (96, 121). In COLEY's time, there was less interest in the cellular changes taking place in toxin-injected tumours; the importance of evoking an acute inflammatory reaction in the tumour was emphasized, however, and the regressing tumour was described as showing the signs of 'inflammatory induration' (125). As mentioned previously, several of the workers of the toxin era suspected the immune system to play a part in the response to toxin therapy, and the intra-tumoral injections were regarded as 'sensitizing the cancer cells'. The early monographs on cancer therapy as practised over a century ago reveal that these early physicians induced severe inflammation at the site of the neoplasm by a variety of irritating or destructive agents such as poultices, vesicants and caustics, while at the same time they stimulated the organism as a whole by the injection of bacterial or plant products at sites remote from the tumour (172). In more recent years, CRILE found that sarcoma 180 implanted on the feet of mice was not suceptible to the action of heat until the tumour was large enough to be the site of an inflammatory reaction (29). The experiments suggested that it was not the heat itself that destroyed the tumour cells, but the inflammatory reaction that followed the heating. CRILE supported this conclusion by the finding that if serotonin, believed to be involved in certain types of inflammation, was injected into a tumour before it was heated, the tumour could be destroyed by less than half the heat exposure normally required (29). Inflammatory exudates destroy neoplastic cells in vitro, as well as causing regression of neoplasms in vivo, as indicated by the work of many investigators (see 125 for refs.). Evidence is accummulating that the cells of the classic inflammatory response (polymorphonuclear leucocytes, macrophages) play an important part in the immune response (24, 110, 182); the inflammatory response can be initiated by immune reactions and by toxins. Most cytotoxic drugs depress not only the immune response but also inflammatory reactions (8). A sharp distinction between the immune response and the inflammatory response is frequently impossible to delineate, and in this perspective, inflammation can be regarded as a part of the entire immune response (8). As mentioned previously, the cells of the reticuloendothelial system (mononuclears, histiocytes, macrophages) are now classed as a major component of the immune system (76).

In the light of current work, it can be seen that elevated temperature and an immune response on the part of the host can each have a deleterious effect on cancer cells. It appears that these two factors are opposite sides of the same coin. The results obtained by the use of COLEY's toxins are irrefutable -- many of these cases were assessed and treated by the most distinguished clinicians and pathologists of the day. The results for the type of advanced cancer treated by the toxins have never been bettered by any other mode of therapy. It is significant that the roles of fever and immunity have not been separated in the response of patients to toxin therapy. Although much progress has been made in the study of the action of heat on cancer cells, it cannot be denied that the treatment of tumours by hyperthermia (local and total body) at present leaves much to be desired. It is difficult to escape the conclusion that more thought should be given to enlisting the aid of the host immune response in the heated patient. This might be achieved by the use of BCG in patients capable of developing delayed hypersensitivity reactions, or by the use of BCG as an adjuvant in combination with suitably inactivated tumour cells (autologous or homologously derived). This latter approach has been reported to evoke a strong anti-tumour response in animal systems (7), and has produced significantly increased survival in patients with advanced malignant melanoma (121). A more recent development is the regression that has been obtained in a spectrum of malignant diseases in man following the induction of a cell-mediated immune reaction by a small organic molecule such as dinitrochlorobenezene, DNCB (96, 159). A combined approach of this type might contain the active components of COLEY's fluid, and it should be possible with out developing knowledge of hyperthermia and tumour immunotherapy to devise a procedure that would yield less inconsistent and capricious results than those of COLEY's era. It must be remembered, however, that in addition to the two components in the host response to COLEY's fluid -- pyrexia and stimulation of the immune system -- the action of the streptococci (or their products) may have served a more direct purpose. For example, there may have been a direct lethal effect of the toxins on the malignant cells, and/or the bacterial enzymes (which include haemolysins, fibrinolysin and hyaluronidase) may have rendered the cells more vulnerable to the host defences. The great disadvantage of COLEY's fluid was that it could not be controlled, either from the point of view of standardizing the toxin injection or predicting the host pyrexia reaction. The extent to which the effects of MBT can be duplicated by the application of controlled hyperthermia and the more refined methods of stimulating the immune response now available awaits investigation.

7. IMPLICATIONS OF CURRENT RESULTS FOR HUMAN THERAPY

A. In vitro Sensitivity of Human Tumours to Hyperthermia (42°C)

The sensitivity of human normal tissues and malignant tumours of different organs to hyperthermia (42°C) was examined *in vitro* by assay systems similar to those used for animal tumours (42). Material for testing was obtained fresh from the operating theatre. It was placed on ice and after removal of any macroscopically necrotic areas, the tissue was cut into thin sections (less than 1 mm^3 in size). These were then allocated at random into aliquots of approximately 100 mg to minimize the effects of tissue or tumour (especially) heterogeneity. For culture, approximately 100 mg tissue was placed in 3 ml Waymouth medium MB 752/1, supplemented with 10% pooled human AB serum in 5 cm diameter plastic Petri dishes. The cultures were placed in a CO_2 incubator at 37.5°C or at 42°C in a CO_2 incubator for 4 h. At the end of this time, respiration and anaerobic glycolysis of the tissue were measured over 6 h at 37.5°C; simultaneously, uptake of radioactive thymidine, uridine and leucine into duplicate cultures over 6 h at 37.5°C was counted. Results were expressed as percentage inhibition or stimulation of these parameters in the heated cultures compared to the control cultures maintained at 37.5°C.

Table 3. Effect of hyperthermia on normal human tissues

Normal tissue		Respiration/Glycolysis after 42°C		
		No change	Increased	Decreased
Breast	(12)	4	6	2
Stomach	(6)	3	2	1
Rectum	(11)	7	2	2
Colon	(8)	5	3	-
Lung	(25)	18	3	4
Skin (breast, prepuce)	(12)	8	2	2
Ovary	(10)	8	2	-
Uterus	(7)	7	-	-
Cervix	(12)	5	3	4
Total	103	65	23	15

Slices of the different tissues were examined over 6 h by Warburg manometry at 37.5°C following a 4 h culture period at 42°C.

Table 3 records the effect of 42°C on over 100 normal tissues in vitro. The material was obtained as a result of surgery for various non-malignant and malignant conditions. Tissue removed from cancerous organs was taken from a site as far removed from the lesion as possible. All tissues used for Table 3 were histologically normal. In the majority of normal tissues, there was little change in O_2 uptake or CO_2 production after hyperthermia. In a minority of samples, 42°C resulted in an increase or a decrease in respiration and/or glycolysis; the extent of the change was usually less than 15% and in none of the tissues examined did it exceed 30%. The inhibition of metabolism was irreversible, however, under the present experimental conditions and no recovery took place over 6 h at 37.5°C after heating.

In contrast, Table 4 shows the percentage of human cancers that responded to elevated temperature by an irreversible decrease of 30% or more in respiration and/or anaerobic glycolysis. With the experimental protocol used, it has been found that total inhibition of respiration or glycolysis is uncommon with human cancers. Instead, when it occurs, inhibition of respiration and/or glycolysis is partial. After exposure to 42°C in culture, inhibition of respiration and/or glycolysis became apparent within 2 h at 37.5°C in Warburg manometers, and over the 6 h period of manometry the degree of inhibition did not increase further. The effect of a pre-incubation period of 20 h in culture at 37.5°C followed by 4 h at 42°C prior to manometry, and also of performing

Table 4. Human cancers - <u>in vitro</u> response to hyperthermia (42°C)

Cancer	No. responding	No. examined	
Lung	14	40	
Breast	20	44	
Stomach	2	14	
Colon	4	8	
Rectum	0	5	
Melanoma	6	10	
Overall response	46	121	(33%)

Legend as for Table 3. 'Responders' are those tumours in which a minimum decrease of 30% occurred in O_2 uptake and/or anaerobic CO_2 production after culture at 42°C.

manometry on tumours immediately after removal from the patient (i.e. without culture) was also investigated. The pattern of response in the tumour slices was similar to that obtained after 4 h in culture; when respiration and/or glycolysis were inhibited by elevated temperature, the decreased gas exchange was measurable within the first 2 h of manometry. With human tumours there was often a decrease in metabolism over 20 h in culture, but this did not affect the response to 42°C in heat-sensitive tumours. In some cancers (especially carcinomas of breast and cervix) O_2 uptake and CO_2 production were so low as to render unfeasible even direct manometry on removal of the material from the patient. Unproductive detailed biochemical studies on such specimens can be avoided by a short (2 h) test period of manometry on receiving the tumour from the operating theatre.

In tumours that were affected by heat, the pattern of response was an individual one for each tumour, and was not related to the tissue of origin (e.g. breast, lung, bowel) or to pathological classification of the neoplasm (e.g. scirrhous carcinoma, adenocarcinoma). Either respiration or glycolysis was inhibited leaving the other parameter unaffected; respiration and glycolysis could both be depressed to equal or unequal degrees; one of these parameters could be decreased and the other increased; and in a small number of cases, both O_2 consumption and CO_2 production were increased following hyperthermia. The majority of tumours belonged to the first two of these categories and the question arose as to what level of inhibition should be regarded as constituting a 'response' to elevated temperature. In tumours in which there was a minimum of 30% inhibition of both O_2 consumption and CO_2 production, there was a corresponding reduction in the uptake of radioactive thymidine, uridine and leucine into the tumour. Over a 6 h culture period following hyperthermia, the inhibition of uptake of these precursors ranged from 40-100%. With tumours exhibiting a minimum 30% decrease of only one parameter (respiration or glycolysis) the effect of 42°C on concomitant precursor uptake was not predictable (sometimes uptake was decreased to various degrees, sometimes it was unaffected).

The relationship between inhibition of metabolism of 30% or more and biological malignancy of the heated cancers was investigated by heterologous transplantation. After heating, the tumours were inoculated into immune-suppressed weanling rats and mice. Animals treated by both cortisone and X-irradiation were found to support the growth of untreated tumours best. A single total body dose of irradiation (100 rads in mice, 150 rads in rats) was administrered 1-4 days prior to tumour implantation; cortisone acetate (60 mg/kg body weight) was injected subcutaneously immediately after implantation, and the steroid dose was repeated on alternate days to a total of four injections (132). For implantation, the tumour was finely diced and introduced into the muscle of the hind leg (approx. 200 mg in rats, 100 mg in mice) by a narrow bore trocar. When adequate material was available, each tumour was inoculated into 3-6 hosts of each species and the animals were examined regularly for tumour growth for at least 12 weeks. Of 22 tumours (10 lung tumours, 8 breast, 2 stomach, 2 colon) in which respiration and glycolysis were inhibited by a minimum of 30% by hyperthermia and that were transplatend into rodent hosts after 4 h in culture at 42°C, none grew. Table 5 shows the results of transplanting 40 control (unheated) tumours into rats and mice. The bowel cancers grew best and produced large tumours up to 2 cm in diameter within 7-14 days. As can be seen from the table, approximately 50% of all tumours transplanted into the rodent hosts grew into measurable and histologically recognizable tumours. Tumours that were heat sensitive when tested in vitro after removal from the patient retained their in vitro response to 42°C after growth in host rats or mice. Cancers that took on transplantation did not grow in all animals inoculated (Table 5). The majority of tumours regressed within 3-4 weeks after implantation.

Table 5. Transplantation of human cancers into pre-conditioned rats and mice

	No. of tumours transplanted		No. of animals used in transplantation	
Cancer	takes	total	takes	total
Lung	8	16	22	121
Breast	3	10	7	42
Stomach	4	6	12	32
Colon	6	8	19	68
Total	21	40	60	263

The tumour material was transplanted into the hind limb of the host animals after a 4 h culture period at 37.5°C.

There is some evidence from animal and human studies that an in vitro inhibition of tumour metabolism in the region of 30% may be significant in relation to tumour response to therapy. DICKSON and SUZANGAR (41) found that the solid Yoshida sarcoma failed to take on transplantation into rats after respiration, glycolysis and nucleic acid and protein precursor uptake had been reduced by 25-30% by a 1 h incubation at 42°C. Tumour regression correlated with as little as 25-30% inhibition of these parameters in vitro by drugs (alkylating agents) over 24 h in culture, while lesser degrees of inhibition were not associated with regression when the appropriate drug dosage was given to the animal. In a review of biochemical approaches to predicting the response of human solid tumours to cytotoxic agents, WOLBERG (186) has pointed out that a clinical response to drug therapy may correlate with an in vitro inhibition of metabolic activity of as little as 30% by the drug.

The pattern of response of human cancers in vitro to 42°C is therefore different to that of animal tumours. A high degree (or total) inhibition of respiration, glycolysis and precursor uptake into nucleic acids and protein does occur but is uncommon. Definitive information as to the significance for the tumour-bearing patient of a 30% or more inhibition of metabolism as produced by hyperthermia in the present experimental approach is unavailable. From the point of view of in vitro predictive sensitivity testing of human cancers, the situation is less satisfactory than with animal tumours. We must bear in mind that heat-sensitive cancers of animals in general require longer to destroy in vivo than in vitro, and that the cancer cell can avail itself of numerous alternative metabolic pathways. It is our current policy, therefore, to regard as heat-sensitive and suitable for thermotherapy only tumours that exhibit in vitro a minimum 30% inhibition of respiration and/or glycolysis with a simultaneous high degree of inhibition (> 5o%) of thymidine, uridine or leucine uptake (42). Determination of such sensitivity is applicable in only a small number of patients referred for thermotherapy. This is because, even with micro-methods of culture and manometry, considerable quantities of tumour are still required. Due to factors such as tumour necrosis, heterogeneity of material and the necessity for pathological examination, adequate cancer tissue for hyperthermia studies is often at a premium.

B. Schedule for Heating Human Tumours in vivo

The response of a heat-sensitive tumour to thermotherapy is governed by the degree of heating and this depends upon both the intra-tumour temperature achieved and the duration of the applied temperature. Our work with animal tumour models has indicated that the optimal intra-tumour and central body temperature to be achieved for systemic hyperthermia is 42°C. Our own clinical experience and that of others (R.T. PETTIGREW, personal communication) substantiates this temperature as being optimal also for human total body thermotherapy; if 42°C is exceeded for a prolonged period then cardiac disturbances (such as arrhythmias), liver dysfunction (e.g. cessation of bile flow), and cerebral derangement (e.g. disorientation, confusion on waking) become likely hazards.

We are then confronted with the question as to how long should be the duration of heating in human therapy. Figure 18 forms the basis of our approach to this problem. In 1940, JOHNSON (90) assembled the results of other workers up to this time on the thermal death times of various types of cancer in vivo and in vitro and his format has been adopted for Figure 18. The unstippled area records the thermal death times of the tumours denoted at bottom right of the panel, and represents data published up to and including 1940. Data obtained post-1940 is contained in the stippled area and refers to tumours listed at top left of the panel. A straight line has been fitted to the pre-1940 in vivo results. The point corresponding to the thermal death time for spontaneous human tumours is based on a review by ROHDENBURG (143) of clinical cases from various sources in which a sustained pyrexia was associated with tumour regression. In these patients, the temperature ranged from 40°C to 40.6°C (104°F-105°F), and the pyrexia lasted for periods varying from 48 to 96 h without remission. The graph indicates the minimum time required at each temperature for tumour destruction. It can be seen that at 42°C a heating period in the region of 20 h is required for tumour destruction. An operating temperature of half a degree less, i.e. 41.5°C, increases the time to the region of 30 h. With temperatures in excess of 42°C, as have been employed with regional perfusion, much shorter heating times are adequate. In the clinical cases of CAVALIERE et al. (19) intra-tumour temperature is

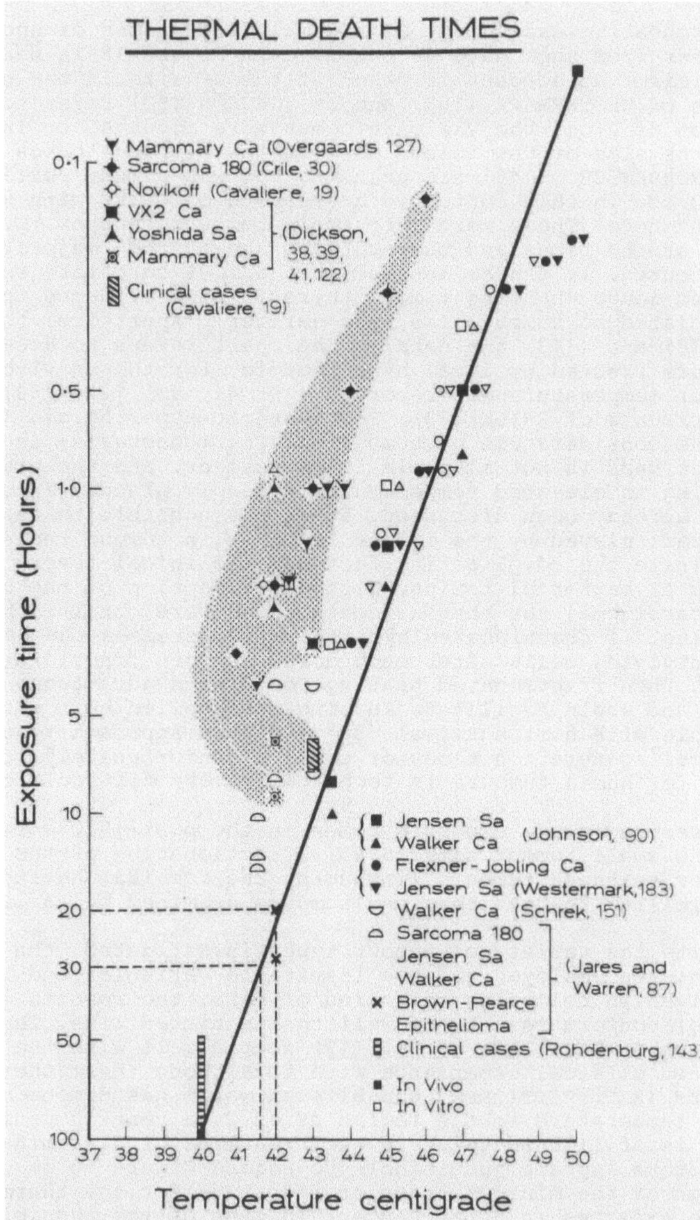

Fig. 18. Thermal death times for animal and human cancers determined in vivo and in vitro by investigators pre-1940 (panel lower right) and post-1940 (panel upper left and stippled area)

published as a range, but it appears that at an average tumour temperature in the region of 43°C, the thermal death time was 6-8 h; at an average tumour temperature in the region of 44°C, massive necrosis and complete tumour regression occurred after a heating period of 3 1/2-4 1/2 h.

The tentative assessment of thermal death times of spontaneous human cancers from such data as compiled in Figure 18 is necessarily open to criticism. No account is taken of tumour size in the chart; the results of WESTERMARK (183) and of JOHNSON (90) refer to tumours in the region of 2 cc, the VX2 carcinomas were about 40 cc in volume (122) and the size of the tumour mass in the clinical cases of CAVALIERE (19) and ROHDENBURG (143) are unknown, but was almost certainly in excess of 40 cc. In this context, the results of CRILE with sarcoma 180 (30) are of note. These were very small tumours (approx. 0.2 cc in volume) and, at the times and temperatures shown, the 'majority' of the tumours were cured. It can be assumed that a 100% cure rate would have required longer heating times; there is also evidence that heating time is related to tumour size (see earlier). Apart from the work reported in (38) and (143) the data on the chart refers to accessible primary tumours treated by local hyperthermia; for the in vivo results intra-tumour temperature was recorded (19, 41, 90, 122, 183) except in the experiments of CRILE (30). For systemic hyperthermia (38, 66), implicit in the consideration of tumour mass is metastasis; in such cases total tumour mass is not amenable to assessment, and the problems of maintaining an elevated temperature in deeply placed, inaccessible lesions have already been discussed. It is not possible to take into account any part played by the immune response in tumour regression, nor to eliminate the complicating factors of clinical fevers such as the effects of bacterial toxins. With the exception of the values for the VX2 carcinoma, the thermal death times were obtained by continous heating. If fractionated hyperthermia increases the rate of proliferation of surviving cells after each dose, as does fractionated radiotherapy (53), then fractionated heating would have advantages for tumour therapy, and would facilitate keeping the applied heat within a range compatible with host survival (38). Such an approach requires knowledge of the cell generation times of tumours; unfortunately, cell kinetic data for human tumours is technically very difficult to obtain.

The seemingly low exposure times in the post-1940 work may therefore be due to small tumour size (S180), fractionation of the heating (VX2) and/or to these tumours (including the Yoshida) having a greater susceptibility to heat than the tumours examined by earlier workers.

Despite the variety of tumour types investigated, the different heating techniques employed and the inevitable variables and imponderables involved in collating this kind of data, the results _in vivo_ (and _in vitro_) conform remarkably well to the fitted line. The recent clinical results of CAVALIERE et al. (19) accord well with the graph. Our own limited clinical experience with total body thermotherapy and that of others (R.T. PETTIGREW, unpublished work) has demonstrated that at a body temperature in the region of 42°C, a heating period of less than 20 h is of limited value in the treatment of disseminated cancer. It therefore appears justifiable to regard Figure 18 as giving some indication of the minimum order of magnitude for the thermal death times to be expected in human cancer. In view of the possible hazards of hyperthermia, it is essential to guard against inadequate heating. Our current belief is that to investigate the potential of this technique and achieve maximum benefit in human cancer by total body hyperthermia, a heating period in the region of 20 h is a reasonable goal (35). To minimize the time spent in the stimulatory temperature range and maximize the lethal effect at 42°C, the hyperthermia should be applied as a single continuous dose, or failing that, as fractionated heating in a minimum number of doses. Until such a strategy is tested in an adequate number of patients with different types of cancer, the place of hyperthermia in the treatment of disseminated malignant disease must remain unknown.

C. Value of Animal Test Systems

The present programme of research on animal tumours was undertaken with a view to deriving basic information and guiding principles for the treatment of human cancer by hyperthermia. It is believed that certain guidelines in relation to the differential sensitivity of cancer cells to heat, the optimal temperature for hyperthermia and the necessity for adequate heating have emerged using animal tumour systems; the clinical results available indicate that such findings have relevance for human cancer therapy.

There is now considerable evidence in favour of the unique nature of human cancer as it occurs in each individual, not only from the point of view of cell metabolism (16) and growth rate (see 56, 91) but also in relation to the inimitable patterns of spread and metastasis that occur in human neoplasia (185). Animal tumours, transplantable or spontaneous, and chosen to provide a reasonably defined growth rate and pattern of behaviour in the host can be studied to obtain reproducible and meaningful results, but the limitations that such standardization imposes must be borne in mind when considering the results in relation to the human situation.

There is also evidence that the host component of the tumour-host relationship differs in animals and man in significant ways. Tumours of the mouse and rat are generally sensitive to heat as indicated in this review; hamster tumours (with the exception of melanomas) are unaffected, as judged by transplantability, by in vitro temperatures as high as 45°C for 1 h, while the Rous sarcoma of fowl is resistant to heating at 50°C (58). The reports of early workers on hyperthermia (144, 183) mention that mice and rats rapidly succumb when body temperature is elevated. In recent detailed studies on rats, DICKSON and ELLIS (36) found that a body temperature of just over 41.6°C represented the 50% lethal dose of heat and few rats survived a temperature of 42°C for 1 h; there was no difference between normal and tumour-bearing rats in susceptibility to heat. This susceptibility to raised temperature circumscribes the use of rodents even for experiments on local hyperthermia, since the blood flow through the heated tumour can easily be large enough in relation to the host's small blood volume to cause a rapid increase in body temperature (36). It can therefore be difficult (depending on anatomical location) to heat even small tumours in rodents for curative periods of time without subjecting the host 'soil' to temperatures above normal. As discussed earlier, elevation of body temperature in rabbits also presents difficulties, thus restricting experiments on total body heating. These animals will not withstand hyperthermia (local or total body) for times in any measure approaching those that at present are required to induce regression of human tumours. VERMEL and KUZNETSOVA (175) have emphasized that, since the thermoregulatory mechanisms in animals are different and less well developed than in humans, the results of hyperthermia experiments on laboratory animals may give little indication of the response when man is the host. A further point in relation to total body heating is that during rapid changes of body temperature the small thermal gradients normally operative between tissues and organs in the animal body can increase considerably (see 38 for refs.). Efficient total body hyperthermia requires stabilization of thermal gradients and the time necessary for this to occur will depend upon the thermoregulatory mechanisms of the host and the heating conditions. The fluctuations in deep temperature encountered in the rabbit during radiant heating may be a reflection of insufficient time to stabilize thermal gradients (38).

It is therefore seen that in seeking animal tumour systems for hyperthermia studies that may be relevant to the human situation, both host and tumour require consideration. The commonly used laboratory animals have very limited aptitude as hosts; suitable and easily transplantable tumours in larger animals, e.g. dogs or pigs, are not available. From the tumour point of view, the different metabolic pattern of response to heat in human cancers compared to animal tumours has led to currently developed in vitro predictability tests being less satisfactory for human cancer than for animal neoplasms.

8. SUMMARY AND PERSPECTIVES

There is now substantial evidence that elevated temperature can have a selective destructive effect on some types of cancer cells. Bacterial toxins (especially those of streptococcal origin) also have an oncolytic effect. The evidence suggests that both these agents bring about alteration and death of susceptible tumour cells in a specific way that involves a response of the host defence mechanisms against the neoplasm. The importance of the competence of the host defences in the response to treatment, and of the inter-play between these defences and therapy in determining prognosis, is receiving increasing recognition in cancer therapy (45, 46, 108).

The application of heat to the primary tumour (local hyperthermia) and to the tumour-bearing host (total body hyperthermia) has enabled some of the complexities of the tumour-host response to treatment to be unravelled and viewed more clearly. This approach has indicated that there are hazards in the use of hyperthermia and that these relate to the effects of heat upon both the tumour and the host. The extent to which the host tissues participate in the response to heat defines local and total body heating as different approaches to cancer therapy.

Although there are similarities in the effects of heat and of bacterial toxins on both the tumour and the host, it seems that pyrexia was not solely responsible for the success of MBT inoculations into cancer patients. The wheel has therefore come full circle and we must view hyperthermia in the role of an adjuvant in cancer therapy, as well as a possible modality of treatment in its own right. It is unlikely that all tumours possess sensitivity to streptococci. Similarly, not all tumours tested are sensitive to elevated temperature (9, 18), and the host probably cannot react immunologically against all cancers. The potential advantages of employing combined therapy in the treatment of cancer, as advocated by other workers (e.g. 109), are evident; with his MBT COLEY may unknowingly have been using a very effective form of such combination therapy.

A major factor preventing 100% tumour cell kill in therapy is the presence of resistant cells in the tumour. For the radiotherapist these constitute the anoxic fraction of the population, and there is now data that such hypoxic cells exist in many human and animal tumours (80). CATER et al. (18) proposed heat in association with radiotherapy as a means of destroying the anoxic radioresistant fraction of the tumour cell population. The ability of heat per se to cure tumours, and the synergism of action between hyperthermia and radiotherapy, supports the value of heat as a means of destroying anoxic as well as oxygenated cancer cells. Recently, ROBINSON and his group have obtained data showing that with increase in temperature the oxygen enhancement ratio decreases and the radiation sensitivity of both oxygenated and anoxic cells is increased. These findings applied to normal and to tumour cells in the mouse, but the effects were more pronounced on the tumour. Thus, hyperthermia in combination with irradiation produced an increase in the therapeutic ratio, which ranged from 1.2 at 40°C to a value greater than 2.0 at 43°C (.J.E. ROBINSON, personal communication).

CRILE (30) induced temporary heat resistance in S91 melanoma and S180 in mice by preheating the tumours for 10-15 min at 44°C the day before the main treatment of 30 min at 44°C which should have cured a large number of the tumours. Several workers have studied the development of thermotolerance in cultured cells and found that most cells surviving a single heat treatment failed to exhibit resistance during subsequent hyperthermic treatment (see 131 for refs.). However, HARRIS (77) induced heat resistance in a line of pig kidney cells by repeated treatments at 47°C, and SELAWRY et al. (153) obtained increased thermotolerance in HeLa, HEp_2 and J96 cells following repeated sublethal heat exposures. At the present time therefore, the induction of heat resistance has not been reported as a problem in tumour thermotherapy, but the possiblity of its occurrence must be borne in mind for inadequately heated cancers.

The prolonged duration of heating at 42°C required for human cancers presents problems of host tolerance in both local and systemic hyperthermia. Agents that would enable a shorter heating period to be employed by potentiating the action of heat e.g. X-rays, or agents that would sensitize the cancer cells and/or render the normal cells resistant to heat are therefore appealing. At the present time, radiotherapy in association with heat would seem most to warrant detailed study in this respect. Recent detailed quantitative studies by ROBINSON et al. (142) on mouse tumours confirm that radiation sensitivity can be greatly increased at elevated temperature. The different points of action of heat and radiation on the cell cycle (184) and on cell biochemistry (130, 184) are further advocates of combination of these two therapies. Hyperthermic cell killing is enhanced by drugs that inhibit RNA synthesis or stimulate DNA and protein synthesis (130). To date, the combination of such drugs and heat has been confined to the in vitro situation (130). The value of this type of approach for animal and human tumours will depend upon the extent to which the normal body cells can be spared while at the same time taking advantage of the selective heat susceptibility of the cancer cells. In this respect, the ability of bacterial toxins to protect the host against radiation damage while at the same time stimulating the reticuloendothelial system, must be kept in mind.

A point for consideration in relation to total body heating is that the elevated temperature resulting from bacterial pyrogens and that of induced hyperthermia involve different mechanisms in the host (115). In febrile illness, it is believed that the bacterial pyrogen produces fever indirectly by interacting with body cells or tissue fluids to produce endogenous pyrogen. This in turn acts directly on thermoregulatory centres to elevate and regulate body temperature at a higher level. In hyperthermia induced by actively heating the body using physical agents, the applied heat over-rides the temperature regulating centres and the physiological response is no longer geared towards reducing heat loss (115). The significance of this difference in physiology in relation to tumour therapy is unknown. An investigation of the effects of induced high temperature on physiological processes must constitute part of our thermotherapy programme. In the context of total body heating in the human, we have little direct information on the susceptibility of various types of internal cancer (e.g. lung, stomach, colon, kidney) to elevated temperature; such relatively inaccessible tumours have not been studied for heat sensitivity in animal systems.

Most transplantable tumours have a faster growth rate than human cancers; studies are needed on the response to heat of slowly growing animal tumours that approximate as closely as possible the primary site, growth rate and other characteristics of human tumours. The availability and suitability of animal model systems in this respect is very limited.

It is therefore seen that by the use of MBT, COLEY opened the lid of a veritable Pandora's box as far as cancer therapy is concerned. The complex nature of the tumour-host response to MBT was only vaguely surmised by the workers of COLEY's era. It is becoming apparent that considerable advances on several fronts are required if we are to benefit fully from the blessings in the box.

Acknowledgements

During the work that has comprised this manuscript, it was my pleasure to have been associated with several dedicated workers. I am deeply indebted to the following for their enterprise and application, and for the privilege of describing their unpublished studies: Mr. D.S. Muckle, F.R.C.S., Mr. D.M. Shah, Dr. M. Suzangar, Dr. M.L. Jasiewicz, Dr. H.A. Ellis and Mr. R. McCoy.

I am obliged to the publishers for permission to reproduce figures as follows: Brit. J. Cancer (Figures 3E, 14); Cancer Research (Figures 2, 4, 8, 10).

The work was supported by the North of England Council of the British Empire Cancer Campaign.

REFERENCES

1. AINSWORTH, E.J., FORBES, F.D.: The effect of Pseudomonas pyrogen on survival of irradiated mice. Radiat. Res. 14, 767 (1961)
2. ANDERVONT, H.B.: The reaction of mice and of various mouse tumours to the injection of bacterial products. Amer. J. Cancer 27, 77 (1936)
3. APITZ, K.: Über Blutungsreaktionen am Impfcarcinom der Maus. Z. Krebsforsch. 40, 50 (1933)
4. VON ARDENNE, M.: Theoretische und Experimentelle Grundlagen der Krebs-Mehrschritt-Therapie. 2nd ed. Berlin. VEB Verlag Volk und Gesundheit, 1971
5. VON ARDENNE, M.: Selective multiphase cancer therapy: conceptual aspects and experimental basis. Advanc. Pharmacol. Chemother. 10, 339 (1972)
6. ASHBURN, L.L.: Spontaneous mammary tumors in mice. Publ. Hlth Rep. (Wash.) 2, 915 (1937)
7. BALDWIN, R.W.: In: Discussion of Evaluation of host immunity in human neoplasia - Part I. Brit. J. Cancer 28, Suppl. 1, 112 (1973)
8. BARLAND, P.: The use of chemotherapeutic agents in the treatment of non-neoplastic diseases. In: Cancer Chemotherapy, 2nd ed. Greenwald, E.S. (ed.). London: Henry Kimpton, Publishers, 1973, p. 406
9. BENDER, E., SCHRAMM, T.: Untersuchungen zur Thermosensibilität von Tumor und Normalzellen in vitro. Acta biol. med. germ. 17, 527 (1966)

10. BIGELOW, W.A., HOPPS, J.A., CALLAGHAN, J.C.: Radiofrequency re-warming in resuscitation from severe hypothermia. Canad. J. med. Sci. 30, 185 (1952)
11. BLACK, M.M.: In: General Discussion of Interaction of Humoral and Cellular Mechanisms in Tumour Immunity. Nat. Cancer Inst. Monogr. 35, 276 (1972)
12. BLOCK, J.B., ZUBROD, C.G.: Adjuvant temperature effects in cancer therapy. Cancer Chemother. Rep. 57, 373 (1973)
13. BROWNING, H.C.: Heterologous and homologous growth of transplant during the course of development of spontaneous mammary tumors in C3H mice. J. nat. Cancer Inst. 8, 173 (1948)
14. BRUNS, P.: Die Heilwirkung des Erysipels auf Geschwulste. Beitr. klin. Chir. 3, 443 (1887)
15. BURKY, E.L.: The production in the rabbit of hypersensitive reactions to lens, rabbit muscle and low ragweed extracts by the action of staphylococcus toxins. J. Allerg 5, 466 (1933-34)
16. BUSCH, H.: Biochemical studies of the tumour cell. In: The Scientific Basis of Surgery. Irvine, W.T. (ed.). London: Churchill, 1965, p. 496
17. BUSCH, W.: Über den Einfluss welchen heftigere Erysipeln zuweilen auf organisierte Neubildungen ausüben. Verhandl. Naturh. Preuss. Rhein. Westpahl. 23, 28 (1866)
18. CATER, D.B., SILVER, I.A., WATKINSON, D.A.: Combined therapy with 220 KV Roentgen and 10 cm microwave heating in rat hepatoma. Acta radiol. Ther. Phys. Biol. 2, 321 (1964)
19. CAVALIERE, R., CIOCATTO, E.C., GIOVANELLA, B.C., HEIDELBERGER, C., JOHNSON, R.O., MARGOTTINI, M., MONDOVI, B., MORICCA, G., ROSSI-FANELLI, A.: Selective heat sensitivity of cancer cells. Cancer (Philad.) 20, 1351 (1967)
20. CHANDLER, J.J., STARK, D.B., ALLEN, C.V., FLETCHER, W.S.: Treatment of cancer by bacterial toxins. Amer. Surg. 31, 443 (1965)
21. CHEN, T.T., HEIDELBERGER, C.: Quantitative studies on the malignant transformation of mouse prostate cells by carcinogenic hydrocarbons in vitro. Int. J. Cancer 4, 166 (1969)
22. CHRISTENSEN, E.A.: Infection and malignant tumours. I. Growth of Brown-Pearce carcinomas in rabbits treated with living or killed haemolytic streptococci. Acta path. microbiol. scand. 46, 285 (1959)
23. CIKES, M., FRIBERG, S. JR., KLEIN, G.: Quantitative studies of antigen expression in cultured murine lymphoma cells. II. Cell-surface antigens in synchronized cultures. J. nat. Cancer Inst. 49, 1607 (1972)
24. COCHRAN, A.J., SPILG, W.G.S., MACKIE, R.M., THOMAS, C.E.: Post-operative depression of tumour directed cell mediated immunity in patients with malignant disease. Brit. med. J. 4, 67 (1972)
25. COLE, D.R., DREYER, B., ROUSSELOT, L.M., TENDLER, M.D.: The radio-protective and anti-tumour effects of mixed bacterial toxins and anthramycin. Amer. J. Roentgenol. 47, 997 (1966)
26. COLEY, W.B.: The treatment of malignant tumours by repeated inoculations of erysipelas with a report of original cases. Amer. J. med. Sci. 105, 487 (1893)
27. COLEY, W.B.: Late results of the treatment of inoperable sarcoma by the mixed toxins of erysipelas and bacillus prodigiosus. Amer. J. Sci. 131, 375 (1906)
28. CRILE, G.: Heat as an adjunct to the treatment of cancer; experimental studies. Cleveland Clin. Quart. 28, 75 (1961)
29. CRILE, G.: Selective destruction of cancers after exposure to heat. Ann. Surg. 156, 404 (1962)
30. CRILE, G.: The effects of heat and radiation of cancers implanted on the feet of mice. Cancer Res. 23, 372 (1963)

31. CRILE, G.: Rationale of simple mastectomy without radiation for crinical stage I cancer of the breast. Surg. Gynec. Obstet. 120, 975 (1965)
32. CRILE, G.: Criticism of conventional methods of treating solid tumours in man. Brit. med. J. 4, 489 (1970)
33. DA COSTA, J.C.: Modern Surgery. 10th ed. Philadelphia: W.B. Saunders, 1931, p. 293
34. DETHLEFSEN, L.A., PREWITT, J.M.S., MENDELSOHN, M.L.: Analysis of tumor growth curves. J. nat. Cancer Inst. 40, 389 (1968)
35. DICKSON, J.A.: Hyperthermia in the treatment of cancer. Cancer Chemother. Rep. 58, 294 (1974)
36. DICKSON, J.A., ELLIS, H.A.: Stimulation of tumour cell dissemination by raised temperature (42°C) in rats with transplanted Yoshida tumours. Nature (Lond.) 248, 354 (1974)
37. DICKSON, J.A., ELLIS, H.A.: The influence of tumor volume and the degree of heating on the response of the solid Yoshida sarcoma to hyperthermia (40-42°C). Cancer Res. 36, 1188 (1976)
38. DICKSON, J.A. MUCKLE, D.S.: Total-body hyperthermia versus primary tumour hyperthermia in the treatment of the rabbit VX2 carcinoma. Cancer Res. 32, 1916 (1972)
39. DICKSON, J.A., SHAH, D.M.: The effects of hyperthermia (42°C) on the biochemistry and growth of a malignant cell line. Europ. J. Cancer 8, 561 (1972)
40. DICKSON, J.A. SUZANGAR, M.: In vitro screening system for human cancer to drugs and hyperthermia (42°C). Brit. J. Cancer 28, 81 (1973); Clin. Oncol. 2, 141 (1976)
41. DICKSON, J.A., SUZANGAR, M.: In vitro - in vivo studies on the susceptibility of the solid Yoshida sarcoma to drugs and hyperthermia (42°C). Cancer Res. 34, 1263 (1974)
42. DICKSON, J.A., SUZANGAR, M.: The in vitro response of human tumours to cytotoxic drugs and hyperthermia (42°C) and its relevance to clinical oncology. In: Biomedical Applications of Organ Culture. Balls, M., and Monnickendam, M.A., Cambridge University Press. 1976, p. 417
43. DONALDSON, S.S., COPPER, R.A., FLETCHTER, W.S.: Effect of Coley's Toxins and irradiation on the A. melanoma # 3 tumor in the golden hamster. Cancer (Philad.) 21, 805 (1968)
44. EDWARDS, J.M.: Malignant melanoma: treatment by endolymphatic radioisotope infusion. Ann. roy. Coll. Surg. Engl. 44, 237 (1969)
45. EILBER, F.R., MORTON, D.L.: Impaired immunologic reactivity and recurrence following cancer surgery. Cancer (Philad.) 25, 362 (1970)
46. FAIRLEY, G.H.: Immunity to malignant disease in man. In: The Scientific Basis of Medicine Ann. Revs. Gilliland, I., and Francis, J. (eds.). London: University of London, Athlone Press, 1971, p. 17
47. FERGUSON, A.B.: Treatment of osteogenic sarcoma. J. Bone J. Surg. 22, 92 (1940)
48. FISHER, B., FISHER, E.R.: Metastases of cancer cells. In: Methods in Cancer Research. Busch, H. (ed.) London and New York: Academic Press, 1967, Vol. I, p. 243
49. FISHER, B., FISHER, E.R.: Studies of metastatic mechanisms employing labelled tumor cells. In: The Proliferation and Spread of Neoplastic Cells. Twenty-first Annual Symposium on Fundamental Cancer Research, Univ. of Texas, M.D. Anderson Hospital & Tumor Inst., Baltimore: Williams and Wilkins, 1968, p. 555
50. FOLEY, E.J.: Antigenic properties of methylcholanthrene-induced tumors in mice of the strain of origin. Cancer Res. 13, 835 (1953)
51. FOULDS, L.: Neoplastic Development. London and New York: Academic Press, 1969, Vol. I

52. FRANKS, L.M.: Structure and biological malignancy of tumors. In: Chemotherapy of Cancer Dissemination and Metastasis. Garattini, S., and Franchi, G., (eds.). New York: Raven Press, 1973, p. 71
53. FRINDEL, E., TUBIANA, M.: Radiobiology and the cell cycle. In: The Cell Cycle and Cancer. Baserga, R., (ed.). New York: Marcel Dekker Inc., 1971, p. 391
54. FUKUI, N.: On a hitherto unknown action of heat ray on testicles. Jap. med. Wld 3, 27 (1923a)
55. FUKUI, N.: Action of body temperature on the testicle. Jap. med. Wld 3, 160 (1923b)
56. GAVOSTO, F., PILERI, A.: Cell cycle of cancer cells in man. In: The Cell Cycle and Cancer. Baserga, R. (ed.). New York: Marcel Dekker Inc., 1971, p. 99
57. GERBER, I.E., BERNHEIM, A.I.: Morphologic study of the reactivity of mouse sarcoma 180 to bacterial filtrates. Arch. Path. (Chicago) 26, 971 (1938)
58. GERICKE, D., CHANDRA, P., ORII, H., WACKER, A.: In vitro thermosensibility of experimental tumours in small animals. Naturwissenschaften 3, 155 (1971)
59. GIOVANELLA, B.C., LOHMAN, W.A., HEIDELBERGER, C.: Effects of elevated temperature and drugs on the viability of L1210 leukemia cells. Cancer Res. 30, 1623 (1970)
60. GIOVANELLA, B.C., MORGAN, A.C., STEHLIN, J.S., WILLIAMS, L.J.: Selective lethal effect of supranormal temperatues on mouse sarcoma cells. Cancer Res. 33, 2568 (1973)
61. GLYNN, L.E., HOLBORROW, E.J.: The production of complete antigens from polysaccharide haptenes by streptococci and other antigens. J. Path. Bact. 64, 775 (1952)
62. GOETZE, O.: Örtliche homogene Überwärmung gesunder und kranker Gliedmassen. Dtsch. Z. Chir. 234, 577 (1932)
63. GOLDENBERG, D.M., LANGNER, M.: Direct and abscopal antitumour action of local hyperthermia. Z. Naturforsch. 26b, 359 (1971)
64. GOLDIN, A.: Animal tumour model: introduction to the rationale of screening tests. In: The Design of Clinical Trials in Cancer Therapy. Staquet, M., (ed.). Brussels: Edition Scientifiques Européennes, 1972, p. 7
65. GOLDIN, A.: Effects of drugs on disseminated tumour. In: Chemotherapy of Cancer Dissemination and Metastasis. Garattini, S., and Franchi, G. (eds.). New York: Raven Press, 1973, p. 341
66. GOTTSCHALK, S.: Zur Behandlung des ulcerirenden inoperablen Cervixcarcinoms. Zbl. Gynäk. 79 (1899)
67. GREENE, H.S.N.: A conception of tumour autonomy based on transplantation studies: A review. Cancer Res. 11, 899 (1951)
68. GREENE, H.S.N., HARVEY, E.K.: The inhibitory influence of a transplanted lymphoma on metastasis. Cancer Res. 20, 1094 (1960)
69. GREENE, H.S.N., HARVEY, E.K.: The relationship between the dissemination of tumour cells and the distribution of metastases. Cancer Res. 24, 799 (1964)
70. GREENE, H.S.N., NEWTON, B.L.: Evolution of cancer of the uterine fundus in the rabbit. Cancer (Philad.) 1, 82 (1948)
71. GREENSTEIN, J.P.: Some biochemical characteristics of morphologically separable cancers. Cancer Res. 16, 641 (1956)
72. GREENWALD, M.D.: Cancer Chemotherapy. 2nd ed. London: Henry Kimpton, 1973
73. GRIFFITHS, J.D., SALSBURY, A.J.: The fate of circulating Walker 256 tumour cells injected intravenously in rats. Brit. J. Cancer 17, 546 (1963)
74. GULLEDGE, A.D.: Biologic factors influencing metastasis of spontaneous mammary gland carcinoma in mice. Proc. Amer. Ass. Cancer Res. 3, 231 (1961)
75. GUTTERMAN, J.: In: Discussion of Diagnosis and Immunotherapy of Cancer. Nat. Cancer Inst. Monogr. 35, 410 (1972)

76. HANNA, M.G., SNODGRASS, M.J., ZWAR, B., RAPP, H.J.: Histopathology of Myobacterium bovis (BCG)-mediated tumor regression. Nat. Cancer Inst. Monogr. 35, 345 (1972)
77. HARRIS, M.: Temperature resistant variants in clonal populations of pig kidney cells. Exp. Cell Res. 46, 301 (1967)
78. HENDERSON, M.A., PETTIGREW, R.T.: Induction of controlled hyperthermia in treatment of cancer. Lancet (1971) 1, 1275
79. HEINE, U., SVEREK, L., KONDRATIK, J., BONAR, R.A.: The behaviour of HeLa-S3 cells under the influence of supranormal temperatures. J. Ultrastruct. Res. 34, 375 (1971)
80. HENK, J.M.: Tumour oxygenation and radiotherapy. In: Modern Trends in Oncology -- Clinical Progress. Raven, R.W., (ed.). London: Butterworth & Co. Ltd., 1973, Vol. I, Pt. 2, p. 217
81. HO, H.S., GUY, A.W., SIGELMANN, R.A., LEHMANN, J.F.: Microwave heating of simulated human limbs by aperture sources. IEEE Trans. Microwave Theory & Techniques 19, 224 (1971)
82. HODGMAN, C.D.: (ed.) Handbook of Chemistry and Physics. 33rd ed. Cleveland, Ohio: Chemical Rubber Pub. Co., 1951, p. 289
83. HOLLCRAFT, J.W., SMITH, W.W.: Endotoxin treatment and X-irradiatio in mice bearing transplanted tumors. J. nat. Cancer Ins. 21, 311 (1958)
84. HOSKINS, J.M., MEYNELL, G.G., SANDERS, F.K.: A comparison of methods for estimating the viable count of a suspension of tumour cells. Exp. Cell Res. 11, 297 (1956)
85. HUGGINS, C., YANG, N.C.: Induction and extinction of mammary cancer. Science 137, 257 (1962)
86. HUMPHREY, J.H.: The fate of antigen and its relationship to the immune response. Antibiot. Chemother. 15, 7 (1969)
87. JARES, J.J., WARREN, S.L.: Physiological effects of radiation. I. A study of the in vitro effect of high fever temperatures upon certai experimental animal tumours. Amer. J. Roentgenol. 41, 685 (1939)
88. JASIEWICZ, M.L., DICKSON, J.A.: Potentiation of the Destructive effect of heat (42°C) on synchronized cancer cells in culture by cell specific antiserum. J. Therm. Biol. 1, 221 (1976)
89. JOHNSON, C.C, GUY, A.W.: Nonionizing electromagnetic wave effects in biological materials and systems. Proc. IEEE 60, 692 (1972)
90. JOHNSON, H.J.: The action of short radio waves on tissues. III. A comparison of the thermal sensitivities of transplantable tumours in vivo and in vitro. Amer. J. Cancer 38, 533 (1940)
91. JOHNSON, R.O., WOLBERG, W.H.: Cellular kinetics and their implications for chemotherapy of solid tumours, especially cancer of the colon. Cancer (Philad.) 28, 208 (1971)
92. KARRER, K.: Doses, schedules and host immunity in animal tumour chemotherapy. In: Pharmacological Principles in Antitumour Chemotherapy. HALL, T.C., and KARRER, K., (eds.). Vienna Academy of Medicine, 1968, p. 119
93. KETCHAM, A.S., KINSEY, D.L., WEXLER, H., MANTEL, N.: The development of spontaneous metastases after the removal of a "primary" tumour. II. Standardization protocol of 5 animal tumors. Cancer (Philad.) 14, 875 (1961)
94. KETCHAM, A.S., WEXLER, H., MANTEL, N.: The effect of removal of a "primary" tumor on the development of spontaneous metastases. I. Development of a standardized experimental technic. Cancer Res. 19, 940 (1959)
95. KLEIN, E.: Introduction: Diagnosis and Immunotherapy of Cancer. Nat. Cancer Inst. Monogr. 35, 331 (1972)
96. KLEIN, E., HOLTERMANN, O.A.: Immunotherapeutic approaches to the management of neoplasms. Nat. Cancer Inst. Monogr. 35, 379 (1972)
97. KOSHIMURA, S., HIRATA, R., SHOIN, S.: On the Streptolysin-synthesizing and anticancer activities of cell-free extract from living hemolytic streptococci. Cancer Chemother. Rep. 13, 107 (1961)

98. LAIRD, A.K.: Dynamics of growths in tumours and in normal organisms. Nat. Cancer Inst. Monogr. 30, 15 (1969)
99. LEMON, A.M., FOLEY, J.F.: Antimetabolite therapy of advanced carcinoma and sarcoma. In: Controversy in Internal Medicine. INGELFINGER, F.J., RELMAN, A.S., and FINLAND, M. (eds.). Philadelphia: Saunders, 1966, p. 575
100. LEWIS, M.R., APTEKMAN, P.M.: Atrophy of tumours caused by strangulation and accompanied by development of tumor immunity in rats. Cancer (Philad.) 5, 411 (1952)
101. LEWIS, M.R., COLE, W.H.: Experimental increase of lung metastases after operative trauma (amputation of limb with tumor). Amer. med. Ass. Arch. Surg. 77, 621 (1958)
102. LIEBELT, A.G., LIEBELT, R.A.: Transplantation of tumors. In: Methods in Cancer Research. Busch, H. (ed.). London and New York: Academic Press, 1967, Vol. I, p. 143
103. LIEBELT, R.A., LIEBELT, A.G., GULLEDGE, A.A., CALVERT, J.: Autoregulation -- normal organ and tumor homeostasis. In: The Proliferation and Spread of Neoplastic Cells, Twenty-first Annual Symposium on Fundamental Cancer Research, The University of Texas M.D. Anderson Hospital and Tumor Institute. Baltimore: Williams and Wilkins Co., 1968, p. 733
104. LIN, J.C., GUY, A.W., KRAFT, G.H.: Microwave selective brain heating. J. Microwave Power 8, 275 (1973)
105. LOVE, R., SORIANO, R.Z., WALSH, R.J.: Effect of hyperthermia on normal and neoplastic cells in vitro. Cancer Res. 30, 1525 (1970)
106. LOVELOCK, J.E., SMITH, A.U.: Heat transfer from and to animals in experimental hypothermia and freezing. Ann. N.Y. Acad. Sci. 80, 487 (1959)
107. MARIE, P., CLUNET, J.: Fréquence des métastases viscérales chez les souris cancéreuses apres ablation chirurgicale de leur tumeur. Bull. Ass. franc. Cancer 3, 19 (1910)
108. MARTIN, D.S., FUGMANN, R.A., HAYWORTH, P.: Surgery, cancer chemotherapy, host defenses and tumour size. J. nat. Cancer Inst. 29, 817 (1962)
109. MATHE, G.: Active immunotherapy. Advanc Cancer Res. 14, 1 (1971)
110. McCLUSKEY, R.T., COHEN, S.: (eds.) Mechanisms of Cell Mediated Immunity. New York: John Wiley & Sons, 1974
111. MENDELSOHN, M.L.: Cell proliferation and tumour growth. In: Cell Proliferation. Lamerton, L.F., and Fry, R.J.M. (eds.). Oxford: Blackwell, 1963, p. 190
112. MENDELSOHN, M.L.: The kinetics of tumor cell proliferation. In: Cellular Radiation Biology. M.D. Anderson Hospital Symposium. Baltimore: Williams and Wilkins, 1965, p. 498
113. MILLER, G.F., KETCHAM, A.S.: Effect of bacterial infection on tumor cell contamination of operative wounds. Surg. Forum 13, 98 (1962)
114. MILLER, T.R., NICHOLSON, J.T.: End results in reticulum cell sarcoma of bone treated by bacterial toxin therapy alone or combined with surgery and/or radiotherapy (47 cases) or with concurrent infection (5 cases). Cancer (Philad.) 27, 524 (1971)
115. MINARD, D., COPMAN, L.: Elevation of body temperature in disease. In: Temperature, its Measurement and Control in Science and Industry. Hardy, J.D. (ed.). New York: Reinhold Pub. Co., 1963, Vol. III, Pt. 3, p. 253
116. MITCHISON, N.A.: Passive transfer of transplantation immunity. Proc. roy. Soc. B, 142, 72 (1954)
117. MONDOVI, B., AGRO, A.F., ROTILIO, G., STROM, R., MORICCA, G., ROSSI-FANELLI, A.: The biochemical mechanism of selective heat sensitivity of cancer cells. II. Studies on nucleic acids and protein synthesis. Europ. J. Cancer 5, 137 (1969)

118. MONDOVI, B., SANTORO, A.S., STROM, R., FAIOLA, R., ROSSI-FANELLI, A.: Increased immunogenicity of Ehrlich ascites cells after heat treatment. Cancer (Philad.) 30, 885 (1972)
119. MONDOVI, B., STROM, R., ROTILIO, G., AGRO, A.F., CAVALIERE, R., ROSSI-FANELLI, A.: The biochemical mechanism of selective heat sensitivity of cancer cells. I. Studies on cellular respiration. Europ. J. Cancer 5, 129 (1969)
120. MORTON, D.L.: In: Discussion of Diagnosis and Immunotherapy of Cancer. Nat. Cancer Inst. Monogr. 35, 412 (1972)
121. MORTON, D.L.: Immunotherapy of human melanomas and sarcomas. Nat. Cancer Inst. Monogr. 35, 375 (1972)
122. MUCKLE, D.S., DICKSON, J.A.: The selective inhibitory effect of hyperthermia on the metabolism and growth of malignant cells. Brit. J. Cancer 25, 771 (1971)
123. MUCKLE, D.S., DICKSON, J.A.: Hyperthermia (42°C) as an adjuvant to radiotherapy and chemotherapy in the treatment of the allogeneic VX2 carcinoma in the rabbit. Brit. J. Cancer 27, 307 (1973)
124. MÜLLER, C.: Therapeutische Erfahrungen an 100 Kombinationen von Röntgen und Hochfrequenz resp. Diathermiebehandlungen bösartiger Neubildungen. Münch. med. Wschr. 59, 1546 (1912)
125. NAUTS, H.C., FOWLER, G.A., BOGATKO, F.H.: A review of the influence of bacterial infections and bacterial products (Coley's toxin) on malignant tumours in men. Acta med. scand. Suppl. 276, 1 (1953)
126. OVERGAARD, K., OKKELS, H.: Über den Einfluss der Wärmbehandlung auf Woods Sarkom. Strahlentherapie 68, 587 (1940)
127. OVERGAARD, K., OVERGAARD, J.: Investigations on the possibility of a thermic tumour therapy - I. Shortwave treatment of a transplanted isologous mouse mammary carcinoma. Europ. J. Cancer 8, 65 (1972a)
128. OVERGAARD, K., OVERGAARD, J.: Investigations on the possibility of a thermic tumour therapy - II. Action of combined heat-roentgen treatment on a transplanted mouse mammary carcinoma. Europ. J. Cancer 8, 573 (1972b)
129. PAGET, S.: The distribution of secondary growths in cancer of the breast. Lancet (1889) 1, 571
130. PALZER, R.J., HEIDELBERGER, C.: Influence of drugs and synchrony on the hyperthermic killing of HeLa cells. Cancer Res. 33, 422 (1973a)
131. PALZER, R.J., HEIDELBERGER, C.: Studies on the quanitative biology of hyperthermic killing of HeLa cells. Cancer Res. 33, 415 (1973b)
132. PAUL, J.: Cell and Tissue Culture, 4th ed., London: E. & S. Livingston, 1970, p. 301
133. PELNER, L.: Host-tumour antagonism. XI. Recapitulation of various aspects: practical applications of this knowledge to the care of patients. J. Amer. Geriat. Soc. 6, 539 (1958)
134. PERCY, J.F.: Heat in the treatment of carcinomas of the uterus. Surg. Gynec. Obstet. 22, 77 (1916)
135. PETTIGREW, R.T., GALT, J.M., LUDGATE, C.M., SMITH, A.N.: Clinical effects of whole-body hyperthermia in advanced malignancy. Brit. med. J. 4, 679 (1974)
136. PHILLIPS, B., GAZET, J.C.: Effect of antilymphocyte serum on the growth of Hep$_2$ and HeLa cells in mice. Nature (Lond.) 220, 1140 (1968)
137. POTTER, V.R., WATANABE, M.: Some biochemical essentials of malignancy: the challenge of diversity. In: Proceedings of the International Conference on Leukemia-Lymphoma. Philadelphia: Lea and Febiger, 1968, p. 33
138. RAO, P.N., ENGELBERG, J.: HeLa cells: effects of temperature on the life cycle. Science 148, 1092 (1965)
139. REILLY, H.C.: Microbiology and cancer therapy: a review. Cancer Res. 13, 821 (1953)

140. REIMANN, S.P., NISHIMURA, E.T.: Attempts at the chemotherapy of cancer. J. Mich. med. Soc. 48, 453 (1949)
141. RINALDINI, L.M.J.: An improved method for the isolation and quantitative cultivation of embryonic cells. Exp. Cell Res. 16, 477 (1959)
142. ROBINSON, J.E., WIZENBURG, M.J., McCREADY, W., SCHELTEMA, J.: Radiation and hyperthermia - a synergistic effect on tumour sensitivity. Proc. 3rd Int. Conf. med. Phys., Sweden, 1972
143. ROHDENBURG, G.L.: Fluctuations in the growth of malignant tumours in man with special reference to spontaneous recession. J. Cancer Res. 3, 193 (1918)
144. ROHDENBURG, G.L., PRIME, F.: Effect of combined radiation and heat on neoplasms. Arch. Surg. (Chicago) 2, 116 (1921)
145. ROMSDAHL, M.D., CHU, E.W., HUME, R., SMITH, R.R.: The time of metastasis and release of circulating tumour cells as determined in an experimental system. Cancer (Philad.) 14, 883 (1961)
146. ROITT, I.M.: Essential Immunology. Oxford: Blackwell, 1971, p. 48
147. ROWETT, H.C.Q.: The Rat as a Small Animal. London: John Murray, 1960, p. 40
148. SAKHAROV, G., ROSSIYSKIY, D.: Essai de traitement du cancer per l'organotherapie. Acta med. URSS 2, 145 (1939)
149. SCHATTEN, W.E.: An experimental study of postoperative tumour metastases. I. Growth of pulmonary metastases following total removal of primary leg tumour. Cancer (Philad.) 11, 455 (1958)
150. SCHATTEN, W.E., KRAMER, W.M.: An experimental study of postoperativ tumor metastases. II. Effects of anaesthesia, operation and cortisone administration on growth of pulmonary metastases. Cancer (Philad.) 11, 460 (1958)
151. SCHREK, R.: A method for counting the viable cells in normal and malignant cell suspensions. Amer. J. Cancer 28, 389 (1936)
152. SELAWRY, O.S., CARLSON, J.C., MOORE, G.E.: Tumour response to ionizing rays at elevated temperatures: A review and discussion. Amer. J. Roentgenol. 80, 833 (1958)
153. SELAWRY, O.S. GOLDSTEIN, M.N., McCORMICK, T.: Hyperthermia in tissue-cultured cells of malignant origin. Cancer Res. 17, 785 (1957)
154. SELIGMAN, A.M., SACK, T.: Chemical alteration of polysaccharide from Serratia marcescens. III. Reaction of polysaccharide with radioactive p-iodo-benzene diazonium chloride and the use of the product in an experimental and clinical study. J. nat. Cancer Inst. 10, 105 (1949).
155. SHEAR, M.J., PERRAULT, A.: Chemical treatment of tumors. IX. Reaction of mice with primary subcutaneous tumors to injection of a haemorrhage producing bacterial polysaccaride. J. nat. Cancer Inst. 4, 461 (1944)
156. SHINGLETON, W.W., BRYAN, F.A., O'QUINN, W.L., KRUEGER, L.C.: Selective heating and cooling of tissue in cancer chemotherapy. Ann. Surg. 156, 408 (1962)
157. SHINGLETON, W.W., PARKER, R.T., MAHALEY, S.: Abdominal perfusion for cancer chemotherapy with hypothermia and hyperthermia. Surgery 50, 260 (1961)
158. SHIMKIN, M.B., ZON, L.: Role of thombocytopenia in haemorrhage produced in Sarcoma 37. J. nat. Cancer Inst. 3, 379 (1943)
159. SJERNSWARD, J., LEVIN, A.: Delayed hypersensitivity-induced regression of human neoplasms. Cancer (Philad.) 28, 628 (1971)
160. SKIPPER, H.: Kinetic considerations associated with therapy of solid tumours. In: The Proliferation and Spread of Neoplastic Cells. Twenty-first Annual Symposium on Fundamental Cancer Research, Univ. of Texas, M.D. Anderson Hospital & Tumor Inst., Baltimore: Williams and Wilkins, 1968, p. 213

161. SKIPPER, H.E.: The cell cycle and chemotherapy of cancer. In: The Cell Cycle and Cancer. Baserga, R., (ed.). New York: Marcel Dekker Inc., 1971, p. 358
162. SKIPPER, H.E., SCHABEL, F.M., WILCOX, W.S.: Experimental evaluation of anticancer agents. XIII. On the criteria and kinetics associated with "curability" of experimental leukemia. Cancer Chemother. Rep. 35, 1 (1964)
163. SMITH, W.W., ALDERMAN, I.M., GILLESPIE, R.E.: Increased survival in irradiated animals treated with bacterial endotoxins. Amer. J. Physiol. 191, 124 (1957)
164. STEEL, G.G., LAMERTON, L.F.: Cell population kinetics and chemotherapy. In: Human Tumour Cell Kinetics. Nat. Cancer Inst. Monogr. 30, 29 (1969)
165. STEHLIN, J.S.: Hyperthermic perfusion with chemotherapy for cancers of the extremities. Surg. Gynec. Obstet. 129, 305 (1969a)
166. STEHLIN, J.S.: Regional perfusion for soft tissue sarcomas of the extremities. In: Neoplasia in Childhood, Twelfth Annual Conference on Neoplasia, University of Texas, M.D. Anderson Hospital and Tumor Institute, Chicago: Year Book Medical Publishers, 1969b, p. 219
167. STETSON, C.A. JR.: The endotoxin properties of Group A streptococci. J. exp. Med. 104, 921 (1956)
168. STEWART, H.L., SNELL, K.C., DUNHAM, L.J., SCHLYEN, S.M.: Transplantable and Transmissible Tumors of Animals. Armed Forces Inst. Path., Washington D.C. 1959
169. STRAUSS, A.A., SAPHIR, O., APPEL, M.: The development of an absolute immunity in experimental animals and a relative immunity in human beings due to a necrosis of malignant tumors. Swiss Med. J. 86, Suppl. 20, 606 (1956); STRAUSS, A.A., APPEL, M., SAPHIR, O., and RABINOVITZ, A.J.: Immunologic resistance to carcinoma produced by electrocoagulation. Surg. Gynec. Obstet. 121, 989 (1965)
170. SUKHIN, M.J.: Treatment of experimental malignant tumours of Krichevsky-Sinelnikov and Jensen with B. prodigiosum. C. R. Acad. Sci. URSS 32, 671 (1941)
171. SUZUKI, K.: Application of heat to cancer chemotherapy. Nagoya J. med. Sci. 30, 1 (1967)
172. TANCHOU, S.: Recherches sur le traitement médical des tumeurs cancéreuses du sein. Ouvrage pratique basé sur trois cents observations (extraits d'un grand nombre d'auteurs) Germer Bailliere, Paris, 1844
173. TYZZER, E.E.: Factors in production and growth of tumour metastases. J. med. Res. 28, 309 (1913)
174. VANWIJCK, R.R., GODRICK, E.A., SMITH, H.G., GOLDWEITZ, J., WILSON, R.E.: Stimulation or suppression of metastases with graded doses of tumor cells. Cancer Res. 31, 1559 (1971)
175. VERMEL, E.M., KUZNETSOVA, L.B.: Hyperthermia in the treatment of malignant diseases. Probl. Oncol. 16, 96 (1970)
176. VIDAL, E.: Travaux de la Deuxieme Conference Internationale pour l'Etude du Cancer. Paris, 1911, p. 160
177. WARD, P.A.: Inflammation. In: Principles of Pathobiology. Lavia, M.F., and Hill, R.B. (eds.). London: Oxford University Press, 1971, p. 96
178. WARREN, S.L.: Preliminary study of the effect of artificial fever upon hopeless tumor cases. Amer. J. Roentgenol. 33, 75 (1935)
179. WEISS, D.W.: In: Discussion of Diagnosis and Immunotherapy of Cancer. Nat. Cancer Inst. Monogr. 35, 359 (1972a)
180. WEISS, D.W.: Nonspecific stimulation and modulation of the immune response and of states of resistance by the Methanol-Extraction Residue Fraction of Tubercle Bacilli. Nat. Cancer Inst. Monogr. 35, 157 (1972b)
181. WEISS, L.: The Cell Periphery Metastasis and Other Contact Phenomena. Amsterdam: North Holland Pub. Co., 1967

182. WEISS, L.: The Cells and Tissues of the Immune System. Englewood Cliffs, N.J.: Prentice Hall, 1972, p. 128
183. WESTERMARK, N.: The effect of heat upon rat tumours. Skand. Arch. Physiol. 52, 257 (1927)
184. WESTRA, A., DEWEY, W.C.: Variation in sensitivity to heat shock during the cell cycle of Chinese hamster cells in vitro. Int. J. Radiat. Biol. 19, 467 (1971)
185. WILLIS, R.A.: The Spread of Tumours in the Human Body, 3rd ed. London: Butterworth, 1973
186. WOLBERG, W.H.: Biochemical approaches to prediction of response in solid tumours. Nat. Cancer Inst. Monogr. 34, 189 (1971)
187. WOLFF, J.: Die Lehre von der Krebskrankheit. Jena, 1907
188. WOODHALL, B., PICKERELL, K.L., GEORGIADE, N.G., MAHALEY, M.S., DUKES, H.T.: Effect of hyperthermia upon cancer chemotherapy - application to external cancer of head and face structures. Ann. Surg. 151, 750 (1960)
189. WYETH, J.A.: Sarcoma cured by acute inflammation. Med. Rec. 57, 125 (1900)
190. ZAHL, P.A., HUTNER, S.H.: Action of bacterial toxins on tumours. II. Effect of sulphanilamide on toxin-induced haemorrhage. Proc. Soc. Biol. (N.Y.) 51, 285 (1942); Action of bacterial toxins on tumours. III. Some biological properties of purified Salmonella typhimurium endotoxin. Proc. Soc. Biol. (N.Y.) 52, 116 (1943); Action of bacterial toxins on tumours. IV. Distribution of tumour-haemorrhage agents among bacterial species. Proc. Soc. Biol. (N.Y.) 52, 364 (1943); Action of bacterial toxins on tumours. VI. Protection against tumour haemorrhage following heterologous immunization. Proc. Soc. Biol. (N.Y.) 54, 187 (1943)
191. ZBAR, B.: Tumor regression mediated by Mycobacterium bovis (Strain BCG). Nat. Cancer Inst. Monogr. 35, 341 (1972)
192. ZWAVELING, A.: Implantation metastases. Chemotherapeutic prophylaxis and tumor growth in an infected milieu. Cancer (Philad.) 15, 790 (1962)
193. ZWEIFACH, B.W., KIVY-ROSENBERG, E., NAGLER, A.L.: Resistance to whole-body X-irradiation in rats made tolerant to bacterial endotoxins. Amer. J. Physiol. 197, 1364 (1959)

ADDENDUM

The burgeoning interest in hyperthermia as an anti-tumour modality is evidenced by the number of publications that have appeared since this chapter was written. Attention is particularly drawn to the following references:

194. Proceedings of the International Symposium on Cancer Therapy by Hyperthermia and Radiation. Baltimore, Md.: Amer. College Radiology Press, 1976
195. Advances in Radiation Biology, Vol. 6. Eds. Lett, J.T. and Adler, H., Acad. Press, 1976. A major portion of this volume is devoted to detailed consideration of experimental and clinical aspects of hyperthermia in cancer
196. Recent data on the influence of tumour population kinetics and pH on the response to heat reported in: Nature 263, 772, 1976. Brit. J. Cancer 34, 262, 1976

Chapter 5
Hyperthermic Treatment of Tumours: Experimental and Clinical Applications

G. MORICCA, R. CAVALIERE, A. CAPUTO, A. BIGOTTI, and F. COLISTRO

1. INTRODUCTION

The first clinical report on the effects of hyperthermia on cancer, the very earliest concerning this action of heat, was furnished by BUSCH (1) in 1866.

Subsequently, a number of clinicians (1, 3, 37, 35, 24) reported on the beneficial effect of high fever or locally applied heat on tumors (36). In 1918 ROHDENBURG (27) reviewed the literature on spontaneous regression of human tumors and found, that out of 166 carcinomas which regressed, 72 had had high fevers, heat applications, or severe infections, and that out of 26 spontaneously regressing sarcomas, 19 had had high fevers or heat treatment or infections. Although only one of these cases was accepted by EHERSON and COLE (12) as an authentic "spontaneous" regression, the same authors arbitrarily excluded all cases reported before 1900, including many of ROHDENBURG's cases.

More recently, NAUTS et al. (22) reported on 30 selected patients with advanced cancer treated with Coley's toxins (unfiltered extracts of Streptococcus erysipelatis and Bacillus prodigiosus), 25 of whom were alive and free from disease more than 10 years after treatment. It is highly significant that all these patients were treated with early batches of the toxin that were highly pyrogenic. Comparable results were never obtained with later and less pyrogenic batches. However, these authors did not ascribe the beneficial results they obtained to high temperatures. Several clinicians also reported beneficial effects of combined heat and X-ray treatment, but it is difficult to establish which part of the effect is to be ascribed to heat and which to X-ray.

WOODHALL et al. (39) infused 20 patients with head and neck cancer with various alkylating agents at temperatures of about 42°C. Of these, two had complete regression of tumors. However, it is impossible to separate the contributions of temperature and chemotherapy to these results. There is also the possibility that the action of the chemotherapeutic agents might have been enhanced by heat according to the mass law. SHINGLETON et al. (30) perfused abdominal tumors at 38°C for 30 min with alkylating agents while reducing the temperature of the surrounding tissues to 31°C, but without remarkable results, thus demonstrating that the thermal gradient does not effectively enhance the action of chemotherapy on tumors. CRILE (6) very aggressively treated four osteogenic sarcomas by surgically exposing the tumor, heating it with diathermy to 60°C, and treating it with radiation, obtaining complete necrotization of the bone and the tumor.

FOWLER (13) reviewed 33 cases of patients with histologically diagnosed inoperable cancers who had spontaneous pyrogenic infection. The tumors disappeared completely and the patients survived from 5 to 44 years. However, eight of these cases also received what was considered to be insufficient radiotherapy. Finally, in 1966 KIRSCH and SCHMIDT (18) reported on 5 of 48 patients with inoperable cancers who were treated with total body hyperthermia in a hot water bath with additional diathermy. One of the five patients died shortly after treatment and the other four showed minor clinical improvement.

From a review of the literature, the action of heat on tumors seems to be evident. In some cases, e.g., in COLEY's patients, the curative effect presumably involved a more complex mechanism, in which, in addition to the action of temperature, either the direct action of bacteria or bacterial products and the response of the host's immune system must have played a role. In other cases, however, the disappearance of tumors in man due to heat application alone has been documented (4). Also, a review of the recent literature indicates the possibility of a useful combination of heat with chemotherapy (31) and radiotherapy (5, 8, 23, 33). Some of these authors have affirmed that the increased effect of radiotherapy when combined with hyperthermia does not depend on the time relationship between radiation and heating. SUIT (33) states that hyperthermia is an "attractive anti-tumor agent because it is capable of producing profound destructive effects on tumor tissue, and sensitizes cells to radiation."

2. EXPERIMENTAL STUDIES

The clinical application of hyperthermia comprises several techniques. The first choice is between whole body and local hyperthermia.

A. Whole-Body Hyperthermia

Whole-body hyperthermia is, theoretically, the selective sensitivity of cancer cells to heat. A series of experiments were performed as follows:[1]

First, whole body heating was carried out on normal dogs. Under general anesthesia, the dog was placed in a cabinet in which forced air ventilation at controlled temperature and humidity was maintained (Fig. 1a).

A cable from a Siemens Ultratherm 608 machine was placed above the animal. Rectal and esophageal temperatures were recorded continuously; muscle temperatures were measured by termistor needle probes every 10 min after switching off the diathermy.

ECG tracings were normal except for simple sinus tachycardia until the temperature rose to an extremely high level (44°C). At this point, rhythmic disorders were observed, which may culminate in ventricular fibrillation. It should also be noted, and this is obvious, that when tachycardia is very pronounced, one can also hear by auscultation pronounced thrills in the right as well as in the left sides of the heart.

Arterial pressure (AP) decreased, but without affecting oxygenation of the parenchymas of the main organs, and diuresis was maintained.

---

[1]The animal experiments were performed with the collaboration of Dr. L. Taglia.

Fig. 1a

Central venous pressure (CVP) was initially reduced, then settling to normal levels, rising dangerously only coincidentally with the above-mentioned disturbances of cardiac rhythm. Breathing was strongly altered, becoming increasingly superficial with a marked increase in frequency probably due to phenomena of thermodispersion.

Second, a 9-year-old bitch suffering from a malignant neoplasm of the breast was treated with whole-body hyperthermia. The animal, whose urine test was normal, and whose azotemia only reached 0.365 mg %, was given general anesthesia, initially with a short-acting barbiturate and, after intubation, maintained by oxygen and halothane.

After 40 min, a rectal body temperature was 44°C, the temperature of the tumor masses between 43 and 43.6°C. This temperature was maintained for about 1 h, after which the animal became markedly dyspneic and showed some cardiac arrhythmia, and was then cooled down gradually. The animal awoke normally from anesthesia, and 8 h later had recovered its normal vitality and appetite. Two weeks later biopsy showed necrotic tissue in which some neoplastic cells suffering from degeneration phenomena were found. Follow-up was continued over 18 months without evidence of recurrence. The neoplastic masses that were not removed at biopsy had remained quiescent and indications were that if complete recovery had not been achieved, at least the neoplastic process had been stopped.

A second whole-body hyperthermia was performed on a 13-year-old male toy schnauzer, suffering from a rectal carcinoma. Although this animal presented with an azotemia of only 0.40%, its urine had a very low specific gravity and contained albumin, erythrocytes, and numerous hyaline casts (chronic interstitial nephritis). After approximately 25 min of hyperthermia, not only was marked tachycardia observed but also a very intense mitral thrill. Hyperthermia was then interrupted, breathing and cardiac frequency rapidly became normal and the mitral thrill disappeared immediately and completely. Everything seemed to be progressing normally when, 2 h after the end of treatment, cardiorespiratory insufficiency developed and the animal died. In recent months we have treated six more bitches (aged 8-13), all suffering from malignant neoplasms of the breast. There have been no adverse incidents, the animals appear to enjoy perfect health, and the tumor masses are markedly reduced.

B. Local Hyperthermia by Perfusion on Dogs

At this stage, we studied the possibility of performing an effective local hyperthermia bearing two purposes in mind:
1. The standardization of a technique for clinical application
2. The checking *in vivo* of *in vitro* results.

With the first purpose in view, we performed hyperthermic perfusion on the hind limbs of dogs. Mongrel adult dogs weighing 15-20 kg were given general anesthesia , the iliac vessels were exposed, and the common iliac artery and vein were cannulated and connected to a Sigmamotor heart-lung machine. A tourniquet was applied and perfusion was carried out with the blood oxygenated through a disk-oxygenator and warmed at the heat exchanger of the Sigmamotor pump before it entered the arterial line of the limb. The first series of perfusions demonstrated that whole blood was unsuited for hyperthermic perfusion due to the high degree of hemolysis and sludging, which was followed by generalized shock with acute renal failure and partial necrosis of the extremities. Thereafter, this complication was avoided by diluting the blood at the beginning of perfusion with an isotonic solution and calibrating the pressure of the arterial pump at values slightly higher than those of the systemic pressure. It was observed that normal canine tissues can tolerate a temperature of 42°C for 2 h, after which damage sets in, probably due to a microcirculatory failure caused by injury to the vessel walls.

Finally, we established the necessity of recording during hyperthermic perfusion the rectal temperature and CVP, in addition to the parameters that are usually recorded in normal surgical procedures. In fact, continuous recording of CVP can supply valuable data on the blood exchange between the extracorporeal and systemic circulations.

Another series of experiments was performed to verify the effectiveness of perfusion as a means of local hyperthermia in the treatment of tumors.

Seven dogs, all suffering from osteosarcomas or from soft-tissue sarcomas in one of their limbs, were treated by hyperthermic perfusion. After cannulation of the appropriate vein and artery (axillary or iliac vessels) and connection to the extracorporeal circulation machine, the temperature of the limb was raised to about 43°C (44°C-45°C at the heat exchangers). It was noted that the limbs showed noticeable edema and that the skin sometimes showed a bronze coloring. In one case, after 12 h, it was observed that the temperature of the limb being treated was much lower than the body temperature of the animal. The limb continued to cool until the 5th day when signs of dry gangrene were observed. The animal was then killed.

In another case, in which the treated limb displayed only a moderate edema, the animal did not recover from anesthesia and died 8 h later in spite of attempts at resuscitation.

In a further five cases, in spite of the fact that there was pronounced edema of the treated limb, the edema gradually disappeared after 1-2 weeks, without complications. In one case, the tumor mass was greatly reduced after treatment, and remained unchanged for 12 months, when it started to grow again. The owner of the dog refused further treatment and the animal was killed.

In the remaining four cases, follow-up was continued for about 12 months. Biopsies were carried out on tumor masses at intervals: rare neoplastic cells were found which always displayed definite degenerative phenomena.

From our experimental data to date it can be concluded that hyperthermic treatment, either of the whole-body type or of selected areas, can be of the utmost importance in the control of neoplasms in dogs.

C. Experimental Models

In order to have more data, we standardized an experimental model which is reliable for small animals. Yoshida solid sarcomas and Regina Elena Institute (IRE) sarcomas were transplanted in the thighs of male Wistar rat. When the tumor reached a volume of about 2 cm^3 7-10 days after inoculation of 2.5×10^7 cells, hyperthermic perfusion of limb and tumor was performed through the iliac artery and vein (Fig. 1b). The equipment used was a small roller-pump water circulating unit equipped with a disk-oxygenator (Fig. 2a). The temperature was 44-45°C at the heat exchanger. Electric thermometers were inserted in the arterial line coming out of the heat exchanger and the muscles of the limbs; several probes were inserted into the tumor mass in order to detect any thermal gradient.

Temperatures were monitored throughout the procedure and the treatment was regarded as being under way when all areas of the tumor had reached 42°C. A tourniquet was applied to avoid the danger of haemoglobinuria and systemic acidosis, which is difficult to control in small animals. At the end of the treatment the tumors were transplanted in normal animals. In some animals the vascular incisions were repaired or by-passed and the tumor was left in situ in order to check its evolution after hyperthermia. Control animals were treated by perfusion at a temperature of 38°C. The results are summarized in Table 1.

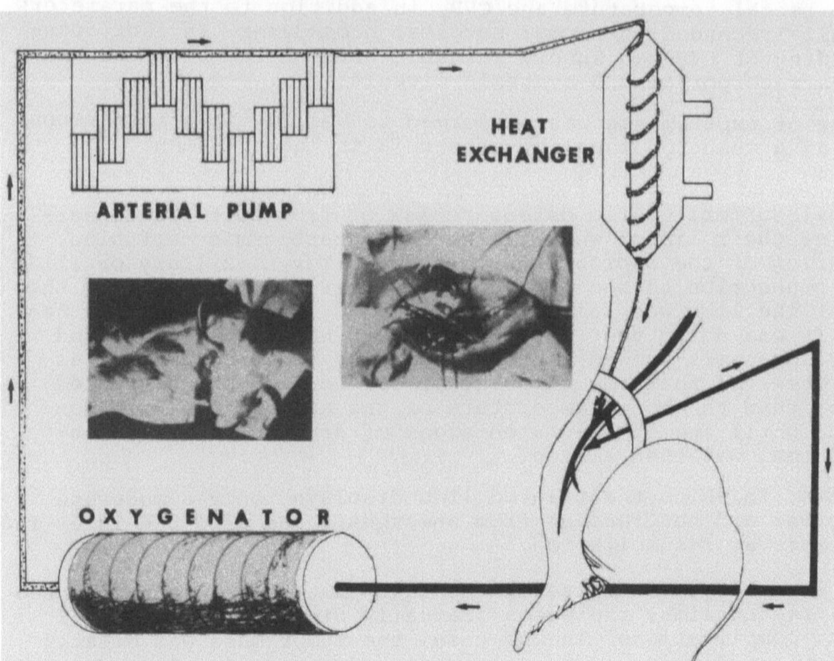

Fig. 1b

Fig. 2a

Table 1. Effect of hyperthermic perfusion on experimental tumors

No. of experiments	Tumor	Site	Hyperthermia (min)	Tumor takes (%)
5	IRE	Thigh	30	0
5	Yoshida	Thigh	45	0
5	Yoshida	Thigh	45	0
5	Yoshida	Kidney	45	0
5	IRE	Kidney	30	0
5	IRE	Insulated Liver	15	0

The experimental results agree with the inhibitory effect of local heating on the growth of experimental tumors observed many years ago by CRILE (5). The inhibitory effects of heat on growth were also observed for Yoshida solid sarcoma, but without affecting the survival of inoculated rats (34). Other authors (17, 38) have not reported positive results, possibly because the heating systems they used were not comparable to ours. The difficulty in achieving and maintaining a high uniform temperature in localized area of the body with the usual systems available (hot air, infrared lamp, ultrasound, diathermy) is well known. We applied heat by perfusing blood which was preheated, diluted, and oxygenated. On the basis of our results it would seem that hyperthermic perfusion enables tissue temperature to be maintained at 42°C homogenously and uniformly over a considerable period of time. Moreover, hyperthermic perfusion seems to be reliable for use in small animals. Our results indicate that the transplantability appears to be inhibited after just 30 min of hyperthermic perfusion. Less precise results were obtained with Yoshida solid sarcoma, but this is a less reliable experimental tumor, possibly because pseudonecrotic or cystic areas easily develop in which it is difficult, sometimes impossible, to achieve hyperthermia by perfusion. In fact, while perfusing Yoshida solid sarcoma, we observed thermal gradients in the tumor "thermal map". This observation may also explain the need for a more prolonged heating time in order to obtain complete regression of the tumor, inhibition of transplantability, and survival of the animals.

These results prompted us to continue hyperthermic treatment in small animals and to devise experimental models for tumor-bearing organs. The left kidneys of male Wistar rats, previously implanted with Yoshida solid sarcomas or IRE sarcoma, were subsequently perfused with prewarmed and diluted blood, with the same equipment used for hyperthermic perfusion of limbs.

Because of the greater lenght of the left renal vessels we always perfused the left kidney, in which the surgical procedures for tumor transplantation and perfusion are simpler. Two techniques were adopted: (1) perfusion carried out through the renal vein and the main trunk of the renal artery with direct suture of these vessels at the end of perfusion, or (2) for a tumor implanted at the upper extremity of the kidney, perfusion through the upper branch of the renal artery with ligature of the branch at the end of perfusion. The latter procedure avoids the complications inherent in a difficult direct suture of a small artery or in a permanent internal bypass. Otherwise, ligature of the upper branch of the renal artery may, perhaps only partially, affect the oxygen supply to the organ with the tumor and thus modify the response of the tumor cells to heat treatment. Perfusion blood flows of about 2 ml/m were employed. Temperatures were recorded by very thin electric thermometers inserted into the tumor and a subcortical area of the kidney. At the heat exchanger the temperature was about 44°C and the oxygenator was previously supplied with blood diluted to 50% with isotonic solution. In control animals perfusions were carried out with the same technique at 38°C. Technically, such an experimental study is difficult to perform on small animals because of the direct surgical exploration of the perfused tumor-bearing organ necessary to check the results. Nevertheless, we obtained some interesting results: the transplantability of IRE sarcoma perfused for 30 min at 42°C was completely inhibited, while regular tumor takes were observed from the tumor perfused as controls at 38°C. It is of interest to note that in 2 of 10 rats in which the perfused kidney was surgically controlled, the tumors regressed completely and the kidneys were functional.

As a final step we carried out hyperthermia of the liver in rats and dogs. Specifically, in rats we performed hyperthermic perfusion of insulated tumor-bearing livers (Fig. 2b) and studied the transplantability of these tumors after treatment. In dogs we studied the technique of hyperthermic perfusion of the liver and its relationship to organ viability and survival of the animals after treatment. For hyperthermic perfusion of insulated tumor-bearing livers we employed a modified procedure of the technique described by KVETINA (20).

In order to reach a uniform temperature of 42°C, it was necessary to perform the perfusion inside a thermostated chamber, thus avoiding elevated heat dispersion phenomena through the surface of the liver. The heat exchanger was maintained at 45°C, the blood diluted and oxygenated, the rate of flows was about 10 ml/m through the portal vein, while venous return, always the locus minoris resistentiae of any perfusion system, was ensured by cannulating the inferior vena cava above the outlet of the suprahepatic veins. In this series of experiments also we showed that the transplantability of heat-perfused tumors was inhibited, while normal tumor takes were obtained from the tumor perfused at 38°C.

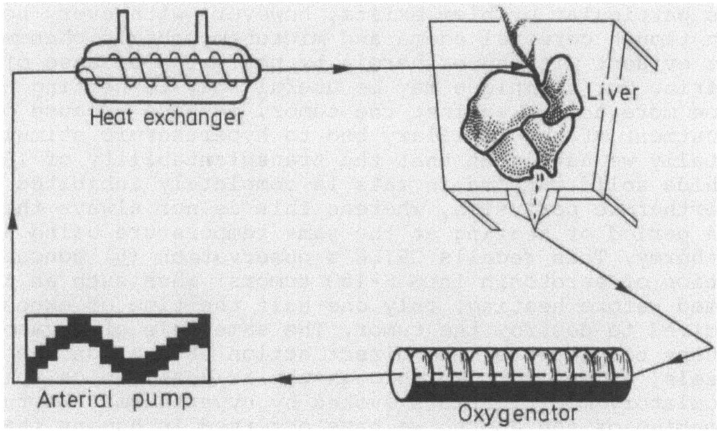

Fig. 2b

Insulation perfusion of the liver in vivo has interesting therapeutic implications, but is difficult to standardize experimentally because of the double circulatory system and the topography of the hepatic veins, which necessitates a complicated thoracolaparatomy, also because of the single pleural cavity of dogs. Therefore we attempted perfusion without insulation in the following sequence: laparatomy and preparation of the hepatic artery through the hepatoduodenal ligament; freeing of the duodenum and insulation of the vena porta with its constituent branches; temporary clamping of the mesenteric superior vein, from which the blood was deviated to the oxygenator and heat exchanger; cannulation through the inferior mesenteric vein of the portal trunk; hyperthermic perfusion of the liver deviating the blood from the upper mesenteric vein, and pumping it after heating into the portal vein system. This perfusion system is effective for heating the liver, but has the disadvantage of introducing warm blood into the systemic circulation. This means a very rapid increase in whole-body temperature in an animal, which does not disperse heat efficiently under general anesthesia. This increase in temperature can be avoided by introducing a heat exchanger in the vena cava or by obtaining the blood from the iliac or femoral artery, reflushing it into a heat exchanger and introducing it into the femoral vein. This is a simple, reliable technique that obviates the need for a pump because the arterial pressure is sufficient to push the blood through such a low-resistance circuit.

At the present state of our knowledge it seems possible to perform hyperthermic perfusion of the liver at 42°C for 2 h without major morphologic or functional impairment. Technically the procedure is not complicated although it does require a complex intra-abdominal preparation. The question is its effectiveness in comparison with whole-body hyperthermia. This question has added importance in view of the large incidence of metastatic neoplastic disease of the liver.

At this point the question arises: Can whole-body hyperthermia be performed by perfusion and can it be more useful than other methods? It is easy to achieve total body heating by perfusion, cannulating the abdominal aorta indirectly through an iliac artery; the heated blood will flow back into the inferior vena cava through an iliac vein, without the need for either oxygenator or arterial pump. In order to maintain the brain temperature below the whole-body temperature, we tried to introduce cold isotonic solutions in the internal carotid arteries via the proximal extremities of insulated external carotid arteries. Since this was found to be ineffective, the problem remains unanswered.

This particular problem exists, however, with every heating technique, even though cerebral edema and microhemorrhagic phenomena might be more evident with hyperthermia by perfusion because of the use of heparin. Our technique may be useful only if heating by perfusion proves to be more active against the tumor, perhaps because of the different adjustment of the capillary bed to hyperthermic stimulation. Experimentally we have seen that the transplantability of IRE sarcoma or Yoshida solid sarcoma in rats is completely inhibited after 30 min of hyperthermic perfusion, whereas this is nor always the case after the same period of heating at the same temperature using a water bath or diathermy. This recalls CRILE's observation (6) concerning the injection of serotonin into S-180 tumors: when such an injection is performed before heating, only one-half the time of exposure to heat is required to destroy the tumor. The same role of a vasoactive drug may perhaps be played by the direct action of heat on the walls of the vessels, or indirectly, through the acidosis which follows the microcirculatory modifications evoked by hyperthermic perfusion. In confirmation of the above, we have observed in humans that the total correction of acidosis and arterovenous shunts during perfusion results in a greater percentage of recurrences.

3. CLINICAL APPLICATION

In beginning our clinical attempts the first problem we were faced with was the choice between whole-body and local hyperthermia.

A. Whole-Body Hyperthermia

We tried whole-body hyperthermia on two patients in order to see the effect of heat on humans with disseminated metastatic cancer. Our first patient was a 25-year-old man (Luigi G.) with disseminated melanoma, hemiplegic due to brain metastases. Under general anesthesia, he was put into a cabinet in which forced air ventilation at controlled temperature and humidity was maintained. The technique was the same described in the experimental studies, except that the head of patient was kept outside of the cabinet and enclosed with the upper part of the neck in a helmet in which cold (4°C) water circulated continuously. A cable from a Siemens 608 Ultratherm diathermy machine was placed above the patient. Rectal and esophageal temperatures were recorded continuously, and the temperatures of the muscles and tumor were measured by thermistor needle probes every 15 min after switching off the diathermy. The first treatment was performed at 41.5°C for 18 h. The immediate results were impressive: the hemiplegia regressed; the patient walked again, although with some difficulty; the cutaneous tumors blistered and many of them regressed during the 2 weeks following treatment. After 1 month all symptoms reappeared, including the hemiplegia. Whole-body hyperthermia was then repeated, and a temperature of 41.8°C was reached and maintained for 23 h. Immediately after heating a syndrome of cerebral edema was instituted and the patient died without recovering consciousness.

Our second patient (Lorenzo M.T.), a 3-year-old boy affected by a rhabdomyosarcoma of the leg with bilateral lung metastases, gave the same results and after an initial brief period of improvement died after the second more intensive whole-body treatment (24 h at 41.5°C). In this case autopsy showed that all the metastatic localizations in the myocardium, pleura, lungs, adrenal glands, etc. were massively necrotic; the brain was edematous with diffuse tiny hemorrhagic areas.

On the basis of the above results, we checked the cerebral temperature in dogs during whole-body hyperthermia and were able to demonstrate that under our experimental conditions there was a thermal gradient between the brain on the one hand and the whole body on the other, with a higher temperature of about 0.8°-1°C in the brain. Our experimental attempts to cool the brain during treatment were unsuccessful. It appears to be more reliable to modify the relationship between the intensity of applied temperature and the time of exposure to heat, especially in view of fractioned repeated exposures. PETTIGREW (25) after an important series of experimental standardization studies, employed a more reliable system of heating the patients, which in addition produces a more uniform temperature throughout the tumor than our method. This is of particular importance because we must bear in mind that hyperthermic treatment is really effective only if the entire tumor cell population is affected by heat. This approach can be modified only when a response to heating by the immune system of the host is established and by the extent to which this happens.

We shall see that, from a general point of view, as far as local heating applied by regional perfusion is concerned, it is possible to answer this question. The opinion of DICKSON (11) on whole-body heating is that the poor results of hyperthermic treatment can be ascribed to a temperature that does not destroy the metastatic cells but on the contrary stimulates their activity. Furthermore the effect of heating may be less marked on metastatic cells than on primary tumor cells, but this view contrasts with our clinical and histologic observation in patients treated with whole-body heating. More accurate investigations are certainly necessary in this field, because of the exceptional importance that whole-body hyperthermic treatment could have in the vast chapter of metastic tumors or tumors conventionally considered untreatable.

B. Local Heating

The perfusion techniques described in the literature so far are not reliable for effective local heating because of the impossibility of achieving and maintaining homgeneous heating throughout the tumor and controlling the temperature effectively. In order to submit human tumors to the action of controlled hyperthermia, we elaborated a new procedure based on the technique of isolation perfusion, introduced by CREECH (9), KREMENTZ (19), and STEHLIN (29) for localized high-dose chemotherapy of cancer of the limbs.

It also seemed reliable because of the availability of amputation as an alternative form of therapy in the event the disease was not well controlled or the limb was damaged irreparably.

4. TECHNIQUE

The first series of hyperthermic perfusions was performed using a Sigmamotor heart-lung machine equipped with a digital pump, a disk-oxygenator previously primed with whole heparinized blood diluted to 50% with an isotonic solution, and a water-circulating unit, maintained at about 47-48°C. The same oxygenator was used for gravity drainage in the venous circuit, in which a filter was inserted to arrest fibrin clots. The patient, under general anesthesia and connected to a mechanical respirator, was positioned on the operating table with a crown of soft gauze under the pelvis or the shoulder in order to avoid bedsores. The whole limb was included in the operative field for the sake of easy access.

Thermistor needle probes applied to the arterial line of the circuit, on the skin, and in the muscles and tumor, were connected to a multichannel thermometer, and perfusion was started. No drugs were perfused. Pump flows and perfusion pressure were previously calibrated. External sources of heat were sometimes added, but this procedure was subsequently discontinued because it produced complications. Also, application of a tourniquet was discontinued after the first cases because of the impairment of nervous sensibility and limb motility. The results obtained in this first series of patients were published in 1967 (4). It must be admitted that many complications arose. Nevertheless, the results of follow-up are encouraging because only one of the patients, who were alive and well in 1967, died of disease, and this after 7 years.

After this first series and because of the severe complications encountered, we improved the technique. The Sigmamotor equipment was replaced with General Electric equipment equipped with a roller pump. We had to standardize the flow equilibration and perfusion pressure during the whole time of perfusion for optimal control of blood exchange between the systemic and perfusion circulations which, if pronounced, would have considerably handicapped limb hyperthermia and the general condition of the patient. This was accomplished by means of a flowmeter inserted into the arterial line of the circuit and a multichannel monitoring system in which systemic and perfusion pressures were recorded, in addition to central venous pressure, ECG, pulse, and respiratory activity. The temperature in the arterial line, before the heated blood entered the artery of the patient, was about 43°C at the beginning of perfusion and about 42.5°C when hyperthermia was achieved. At the heat exchanger the temperature was about 1-1.5°C higher and this temperature was maintained by an independent water-circulating thermostated unit. The raising and maintenance of the temperature was also accomplished by wrapping the limb in warm rubber blankets. Another important modification in the technique concerned the temperature recording system which consisted of two multichannel thermometers with several themistors inserted into various areas of the tumor. Further findings can be summarized as follows. The patients were not heparinized other than through the perfusion circuit, in which a beta-blocking agent was injected at the beginning of perfusion to reduce the reaction to heat at the microcirculatory level with arterovenous shunts and tissue hypoxia originating from exclusion of the capillary bed. Mannitol was slowly but continuously infused into the patient for some days until the urinary findings became normal. Transfusions and fluid and electrolytes were administered as needed; hydrocortisone was avoided in intra- and postoperative care.

Mortality and morbidity were markedly reduced after these modifications to the technique. No acute renal failures, secondary shock, or arterial ruptures were observed, even when the perfused limb suffered postoperatively from swelling, and sometimes blistering, with venous stasis.

A problem of great interest is the choice of solution for use in washing the circuit and limb after perfusion. First we used saline and isotonic solutions, which worsened the condition of the perfused limb, probably because of its acidifying effect.

Therefore we tried to wash the circuit with 3.6% hydroxymethylaminomethane diluted in isotonic solution or added to 1.6% sodium bicarbonate. The results were satisfactory with regard to the postoperative condition of the limb, but at the same time it was disappointing to observe an increased incidence of recurrence. This observation suggested that inflammation played an important part in the action of heat.

5. CASE REPORT

From October 1964 to December 1974 we treated 111 patients for tumor of the limb: 5 squamous cell carcinomas, 27 osteogenic sarcomas, 28 sarcomas other than osteogenic, and 51 melanomas.

Table 2 shows the results obtained in patients treated for squamous cell carcinomas. The only death was caused by myocardial infarct, 13 months after treatment. All these patients needed amputation: in the first case we treated, no remarkable immediate response was observed and the microscopic response was not very intense. The patient refused other therapy and was lost to immediate follow-up. However, he returned to our observation almost 3 years later; the tumor was growing, but the axillary nodes were not enlarged. He refused further perfusion, and therefore he was amputated and followed up without carrying out axillary node dissection. He is now free of disease. The second patient was treated for a large ulcerated squamous cell carcinoma of the foot. The temperature, at that time measured with one thermistor in the tumor, seemed to be effective. The tumor regressed rapidely and a biopsy taken 7 days later showed an almost complete necrosis. Nevertheless the tumor recurred after 1 month and amputation was required. Two further cases of squamous cell carcinoma were extensively ulcerated tumors; the first was amputated at the upper third of the thight because of nonfunctional limb, probably due to the very high temperature necessary to maintain 42°C on the ulcerated surface of the tumor. The second patient amputated because of reappearance of the tumor. These results convinced us of the difficulty of obtaining a really effective treatment by perfusing the ulcerated surface of the carcinoma without major damage to the limb. On the other hand, squamous cell carcinoma is susceptible of cure without perfusion.

The only implication which arises from these cases concerns the first patient, amputated 3 years after recurrence and cured presumably because of the immunoresponse, which may have prevented metastasis of the node.

We treated 55 patients with sarcoma. We regard as particularly interesting the hyperthermic treatment of osteogenic sarcoma, because of its

Table 2. Squamous cell carcinoma

Case	Site	Treatment	Subsequent therapy	Distant results	Elapsed time
1	Hand	Hyperthermic perfusion	Amputation 3 years later for recurrence	N.E.D.	10 years
2	Foot	Hyperthermic perfusion	Amputation 1 month later	N.E.D.	8 years 9 months
3	Thigh and upper leg	Hyperthermic perfusion	Amputation 1 month later	N.E.D.	5 years
4	Hand	Hyperthermic perfusion	Amputation 1 month later	N.E.D.	3 years
5	Thigh and leg	Hyperthermic perfusion	Amputation for non-functional limb	D.O.C.	13 months

N.E.D. = No evidence of disease.
D.O.C. = Died other cause.

immunogenicity and the relatively poor responses obtained by conventional therapy and perfusion chemotherapy. A total of 27 patients were treated for osteogenic sarcoma: three died postoperatively and are not evaluated. We treated the first 9 patients by hyperthermic perfusion (Table 3). In all cases there was a good immediate response to treatment. One patient had a recurrence 6 months later (case 4) and was treated by radiation. He survived 3 years. One patient was amputated because of a nonfunctional limb, another because of recurrence of the tumor some time after an arterial graft and a fourth patient because of recurrence of the tumor 4 months after hyperthermic perfusion.

Table 3. Osteogenic sarcoma

Case	Site	Treatment	Subsequent therapy S: A[a]	Distant results	Elapsed time
1	Tibia	Hyperthermic perfusion	Amputation at time of perfusion	N.E.D.	7 years 2 months
2[a]	Femur	Hyperthermic perfusion	None	N.E.D.	4 years 7 months
3	Femur	Hyperthermic perfusion	Amputation for recurrence 4 months later	N.E.D.	4 years
4	Femur	Hyperthermic perfusion	Radiotherapy for recurrence 6 months later	D.O.D.	3 years
5	Tibia	Hyperthermic perfusion	Amputation of nonfunctional limb 2 months later	D.O.D.	3 years
6	Tibia	Hyperthermic perfusion	Radiotherapy; amputation; chemotherapy	D.O.D.	3 years
7	Tibia	Hyperthermic perfusion	Amputation for recurrence	D.O.D.	3 years
8	Tibia	Hyperthermic perfusion	Amputation for recurrence 3 months later	D.O.D.	2 years 7 months
9	Humerus	Hyperthermic perfusion	Radiotherapy	D.O.D.	1 year

[a]Hyperthermic perfusion alone or followed by other treatment when necessary.
[b]See Figs. 7-12.
N.E.D. = No evidence of disease.
D.O.D. = Died of disease.

Two major considerations arise from this first series of cases. The first concerns the difficulty of immediate follow-up, unreliability of the radiographic pattern for this purpose, and the problem of multiple biopsies. The second concerns the particular vascular situation of osteogenic sarcoma, in which it is hard to achieve and maintain uniform hyperthermia, as we demonstrated later by using many thermometer probes inserted into the tumor. In any event it is possible, after assessing this first series of cases, to conclude that osteogenic sarcoma is affected by heat, as was shown by the fact that three of eight patients, one without other therapy, were respectively, free of disease, 4, 5, and 7 years after treatment.

In addition, we evaluated the possibility of using hyperthermic perfusion in combination with the systematic performance of minor extirpative surgery. In order to utilize the possible immunoresponse to regional heating we decided to carry out the amputation some weeks after the hyperthermic perfusion.

We treated 12 patients with hyperthermic perfusion systematically followed by amputation 4 weeks later. In some instances hyperthermic perfusion was repeated before the amputation (Table 4). It is noteworthy that in two cases no tumor was found histologically. This is confirmation that heat treatment by perfusion can be effective in human osteogenic sarcoma. Moreover, a comparison of the results shown in Tables 3 and 4 indicates that hyperthermic perfusion followed by amputation some weeks later may be the preferred therapeutic management, probably because this interval of time enhances the immunoresponse which is directly or indirectly caused by heat. We believe in this new approach to the problem of the curability of osteogenic sarcoma, even though it is too early to draw definite conclusions. It is our opinion that 6 osteogenic sarcomas out of 12 which may be defined as "controlled" require this kind of intervention.

Obviously our data should be confirmed by further trials and be evaluated statistically. This had not been possible yet because of the distribution of the patients treated during the past years. Nevertheless, in four of the six patients who are free of disease the treatment was performed more than 5 years ago (Fig. 3).

Three patients were treated for osteogenic sarcoma by antiblastic hyperthermic perfusion. As chemotherapeutic agents melphalan and/or dactinomycin were used. All of these patients were amputated after a good immediate response. One died of disease during the first year after treatment. Of the remaining two, one is free of disease after 1 year and the other is living with bilateral lung metastases, which are improving with chemotherapy.

We treated 31 patients for sarcoma other than osteogenic. Three patients died postoperatively and are not evaluated. In these cases hyperthermic treatment was always used in combination with major or minor surgical procedures (see Table 5). The results obtained in this group of patients

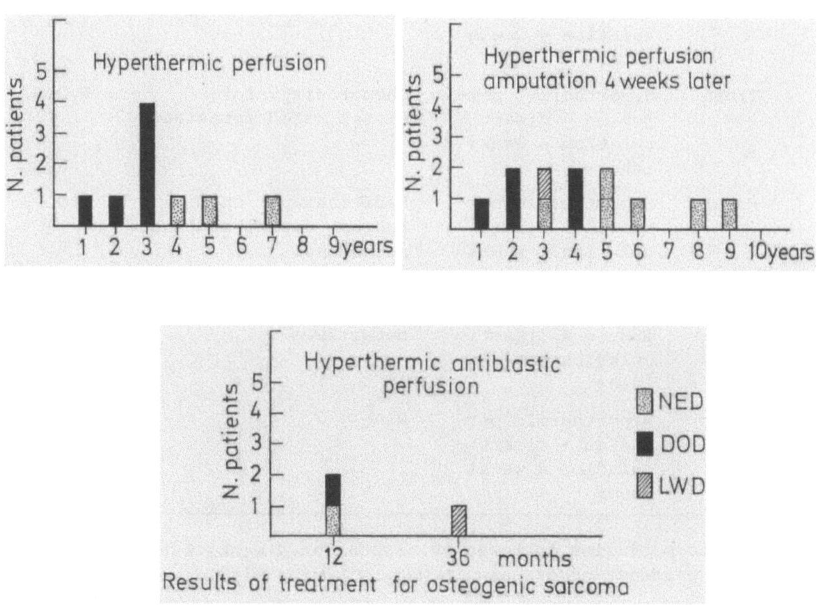

Results of treatment for osteogenic sarcoma

Fig. 3

Table 4. Osteogenic sarcoma

Case	Site	Treatment	Subsequent therapy B[a]	Distant results	Elapsed time
1	Tibia	Hyperthermic perfusion + amputation 4 weeks later	None	N.E.D.	8 years 6 months
2	Femur	Hyperthermic perfusion	Hyperthermic perfusion + disarticulation 4 weeks later	N.E.D.	7 years 10 months
3	Femur	Hyperthermic perfusion + amputation 4 weeks later	None	N.E.D.	6 years
4	Femur and tibia	Hyperthermic perfusion	Hyperthermic perfusion 4 weeks later; hyperthermic perfusion and amputation 4 weeks later; no tumor histologically	N.E.D.	5 years
5	Femur	Hyperthermic perfusion + disarticulation 4 weeks later	None	N.E.D.	4 years 7 months
6	Foot	Hyperthermic perfusion	Hyperthermic perfusion + amputation 4 weeks later	D.O.D.	4 years
7	Femur	Hyperthermic perfusion + amputation 4 weeks later	None	N.E.D.	3 years
8	Femur	Hyperthermic perfusion + disarticulation 4 weeks later	Chemotherapy for lungs metastases	L.W.D.	3 years
9	Tibia	Hyperthermic perfusion + disarticulation 4 weeks later	Chemotherapy for disseminated metastases	D.O.D.	2 years 7 months
10	Femur	Hyperthermic perfusion + disarticulation 6 weeks later	Radiotherapy; chemotherapy for disseminated metastases	D.O.D.	2 years 2 months
11	Femur	Hyperthermic perfusion + disarticulation 4 weeks later	Radiotherapy for lung metastases	D.O.D.	2 years
12	Femur	Hyperthermic perfusion + disarticulation 4 weeks later	None	D.O.D.	1 year

[a]Hyperthermic perfusion followed by amputation 4 weeks later.
N.E.D. = No evidence of disease. L.W.D. = Living with disease. D.O.D. = Died of disease.

Table 5. Sarcoma other than osteogenic

Treatment	No. of patients
Hyperthermic perfusion and excision	11
Hyperthermic antiblastic perfusion and amputation	9
Hyperthermic perfusion and excision or amputation	8

are summarized in Figure 4. It must be emphasized that the choice between simple excision and amputation was made by evaluating the histological type of tumor, its size, and localization. All these tumors were recurrent and previously treated by surgery and in some instances also by radiation; two were metastatic - one from the uterus and the other from the ethmoidal sinus. The latter was the first patient to be perfused. Probably because of a mistake in calibrating the arterial pump, his arm swelled massively after treatment and amputation was necessary a day later. A biopsy taken during this operation showed that the tumor was present in the stump. However, the patient improved without other therapy and is now free of disease more than 10 years after treatment.

A comparative evaluation of the various combinations of surgical excision, amputation, and hyperthermic perfusion in the treatment of soft tissues sarcoma does not reveal significant advantages of one over the other.

The immediate results obtained by means of antiblastic hyperthermic perfusion seem at present more positive than those obtained by means of hyperthermia without chemotherapy, but not enough time has elapsed to enable us to form a conclusive opinion on this subject.

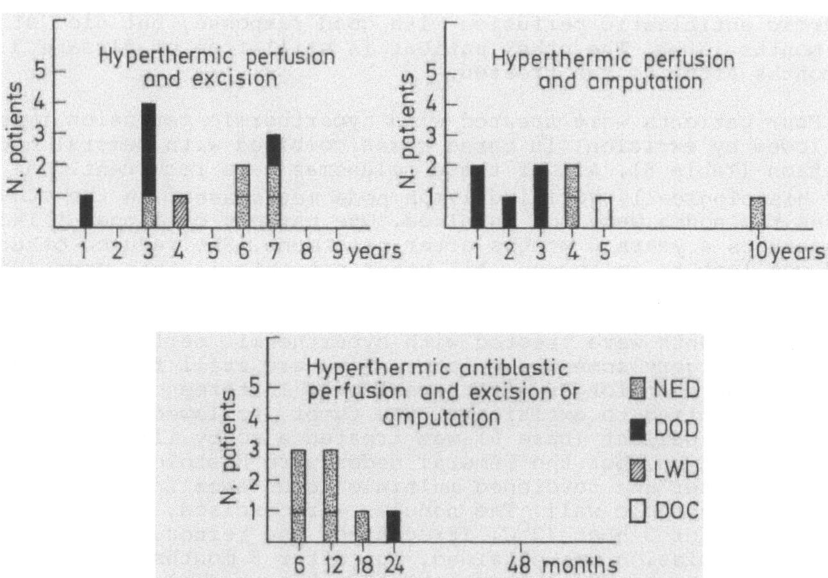

Fig. 4 Results of treatment for soft tissues sacroma

We treated a total of 51 patients affected by melanoma, but only 45 cases are considered here because 6 patients died after treatment. Eleven cases of recurrent melanoma (Table 6) were treated with hyperthermic perfusion alone. Six patients are now free of disease, two of them 10 years after perfusion and the others, respectively, 8 years 2 months, 6 years 10 months, 5 years, and 3 years 1 month after treatment. One patient died 7 years after treatment.

A comparative evaluation of the various combinations of surgical excision, amputation, and hyperthermic perfusion in the treatment of soft tissue sarcoma does not reveal significant advantages of one over the other.

The immediate results obtained by means of antiblastic hyperthermic perfusion seem at present more positive than those obtained by means of hyperthermia without chemotherapy, but not enough time has elapsed to enable us to form a conclusive opinion on this subject.

We treated a total of 51 patients affected by melanoma, but only 45 cases are considered here because 6 died after treatment. Eleven cases of recurrent melanoma (Table 6) were treated with hyperthermic perfusion alone. Six patients are now free of disease, two of them 10 years after perfusion and the others, respectively, 8 years 2 months, 6 years 10 months, 5 years, and 3 years 1 month after treatment. One patient died of disease 7 years after treatment. However, she had a distant metastasis on the thoracic wall at the time of treatment.

In order to evaluate systematically the possible therapeutic combinations of hyperthermic perfusion, four other groups of patients were treated according to the following pattern:

1. Two patients (Table 7) were treated for a primitive melanoma with hyperthermic perfusion after a wide bioptic excision. In both patients, probably because of the extent of the surgical excision, hyperthermia had to be incomplete and did not reach the required heat values. One patient developed metastases after 8 months. He was treated by hyperthermic antiblastic perfusion with good response, but died of disease 11 months later. The other patient is still free of disease 3 years 4 months after he was treated.

2. Four patients were treated with hyperthermic perfusion immediately followed by excision, in three cases combined with femoral node dissection (Table 8). All of these melanomas were recurrent. One patient had histologically verified lymph node metastases; in the other two cases the nodes were not involved. One patient developed iliac node metastases 4 years 8 months after treatment. She refused other therapy, and was lost to follow-up. All other patients in this group died within 2 1/2 years.

3. Ten patients were treated with hyperthermic perfusion (Table 9) followed by surgery some weeks later. Five are still free of disease 4-7 years and even longer after treatment. In three of these patients surgery was limited to excision of the tumor, followed in one case by skin grafts. One patient (case 6) was treated also by iliac and femoral lymph node dissection, but the femoral nodes were histologically involved. One year later she developed multiple metastases in the breast and posterior thoracic wall. The nodules were excised, the cells insulated and heated for 3 h at $42^{\circ}C$, irradiated, and reinoculated subcutaneously. A remission was obtained, but after 8 months she developed disseminated metastases in the limbs and skin, and was treated again with heated irradiated cells and chemotherapy. She died of disease 3 years and 1 month after the first treatment. These results are better than

Table 6. Melanoma: Hyperthermic perfusion alone

Case	Site	TNM classification	Time of hyperthermia	Subsequent therapy	Distant results	Elapsed time
1	Heel	$T_3N_OM_O$	2 h 45 min 41.5-44°C	None	N.E.D.	10 years
2	Leg	$T_ON_1M_1$	4 h 41.5-42.5°C	Hyperthermic perfusion	N.E.D.	9 years 11 months
3	Foot	$T_2N_1M_O$	2 h 41.3-43°C	None	N.E.D.	8 years 2 months
4	Leg	$T_3N_1M_1$	4 h 42-43.5°C	Amputation for vascular insuf- fiency 12 months after perfusion	D.O.D.	7 years
5	Thigh	$T_3N_OM_O$	4 h 41.8-42.5°C	None	N.E.D.	6 years 10 months
6[a]	Foot	$T_2N_OM_O$	3 h 42-42.5°C	None	N.E.D.	5 years
7	Entire leg	$T_ON_3M_1$	4 h 41.8-42.3°C	None	N.E.D.	3 years 1 month
8	Leg	$T_2N_1M_1$	2 h 42°C	Patient refused further therapy	Lost to follow-up	1 year 10 months
9	Unknown primary enlarged axillary nodes	$T_ON_3M_O$	4 h 30 min 42.1-43°C	None	D.O.D.	9 months
10	Arm	$T_3N_3M_1$	4 h 15 min 42-44°C	None	D.O.D.	5 months
11	Entire	$T_ON_2M_1$	3 h 40 min 42-42.8°C	None	D.O.D.	3 months

[a]See Figs. 36-49.
N.E.D. = No evidence of disease.
D.O.D. = Died of disease.

Table 7. Melanoma: Hyperthermic perfusion following biopsy

Case	Site	TNM classification	Time of hyperthermia	Subsequent therapy	Distant results	Elapsed time
1	Leg	$T_2N_OM_O$	Incomplete hyperthermia	None	N.E.D.	3 years 4 months
2	Leg	$T_2N_2M_O$	Incomplete hyperthermia	Antiblastic hyperthermic perfusion	D.O.D.	1 year 7 months

N.E.D. = No evidence of disease.
D.O.D. = Died of disease.

Table 8. Melanoma: Hyperthermic perfusion immediately followed by excision

Case	Site	TNM classification	Time of hyperthermia	Subsequent therapy	Distant results	Elapsed time
1	Leg	$T_3N_OM_1$	4 h 10 min 41.8-42.9°C	None	D.O.D.	2 years 6 months
2	Leg	$T_ON_1M_1$	3 h 35 min 41.5-42.4°C	Skin graft	Refused therapy 4 years 8 months later for enlarged iliac nodes	Lost to follow-up
3	Foot lateral surface	$T_3N_1M_1$	4 h 42.5-42.4°C	Refused further therapy	D.O.D.	2 years 1 month
4	Knee	$T_3N_3M_O$	4 h 41.8-43.6°C	None	D.O.D.	2 years

D.O.D. = Died of disease.

those reported previously, probably because of the interval of time between perfusion and excision. This confirms that enhancement of the immunoresponse is important in the mechanism of heat action. One patient (case 3), a 20-year-old woman, had a serious hemorrhage postoperatively. It was treated surgically with arterial prosthesis because of the multiple ruptures of the iliac artery, probably due to infection caused by protracted drainage. After another sudden hemorrhage the common iliac artery was ligated. She is free of disease and her limb is functioning well, 5 years 6 months after treatment.

4. Eighteen other patients were treated with two different combinations of hyperthermia and chemotherapy. Six patients (Table 10) were treated with hyperthermic perfusion; the last perfusion hour was carried out after applying a tourniquet and administering an antiblastic drug. Only one patient is still alive and well. He was perfused with glyceraldehyde (3 g). The effect of such combined therapy were disastrous for the limb: it swelled enormously and blistered, and a secondary venous insufficiency occurred on the third day, progressing until gangrene set in despite cutaneous débridements and fasciotomy. The limb was amputated 10 days after treatment, and the patient recovered slowly. Glyceraldehyde was no longer administered in combination with hyperthermic perfusion. Subsequently we used melphalan, which in this group of patients was administered at a dose of 0.8 mg/kg body weight. The results were not satisfactory. We then tried the combination of hyperthermic perfusion with chemotherapy according to STEHLIN (31), i.e. performing hyperthermic perfusion with a tourniquet from the very beginning, and at the same time injecting the drug into the arterial line. The hyperthermic antiblastic perfusions were carried out at 42-43°C for 90-120 min. We treated 12 patients with this technique. (Another patient, treated with hyperthermic antiblastic perfusion 8 months after a wide bioptic excision followed by hyperthermic perfusion, is included in Table 6). Six of these patients are at this writing free of disease, but the longest interval of time elapsing since treatment is 3 years 10 months. Therefore, no conclusions can be drawn yet.

Table 9. Melanoma: Hyperthermic perfusion followed by surgery 4 weeks later

Case	Site	TNM classification	Time of hyperthermia	Subsequent therapy	Distant results	Elapsed time
1	Foot - plantar surface	$T_2N_1M_0$	4 h 41.5-42°C	None	N.E.D.	7 years 1 month
2	Leg - below knee	$T_3N_0M_1$	4 h 42-43°C	Skin graft	N.E.D.	7 years 1 month
3	Foot - lateral plantar surface	$T_2N_1M_0$	4 h 42-43.5°C	None	N.E.D.	5 years 6 months
4	Fourth Toe	$T_2N_1M_0$	4 h 45 min 39-41.3°C	None	N.E.D.	4 years 2 months
5	Foot	$T_2N_3M_0$	2 h 40°C	None	N.E.D.	4 years 1 month
6	Leg	$T_2N_3M_0$	3 h 30 min 40.2-42.8°C	Treatment by irradiated cells	D.O.D.	3 years 1 month
7	Leg	$T_3N_1M_1$	2 h 30 min 42-43.5°C	None	D.O.D.	2 years 7 months
8	Leg	$T_2N_3M_0$	3 h 41.6-43°C	None	D.O.D.	1 year 2 months
9	Entire leg	$T_0N_0M_1$	4 h 41.5-43.8°C	None	D.O.D.	9 months
10	Entire leg	$T_3N_1M_1$	3 h 40 min 41.8-42.5°C	Hyperthermic perfusion 2 months later at 42°C for 2 h and dis-articulation after 4 weeks	L.W.D.	2 years 8 months

N.E.D. = No evidence of disease.
D.O.D. = Died of disease.
L.W.D. = Living with disease.

To sum up, we tried all possible perfusion treatments which hyperthermia admits. Figures 7-49 present the pathological findings. The patients who were treated with hyperthermic perfusion alone or in combination with surgical excision were followed up for significant periods of time. It is our opinion that hyperthermic perfusion can be an effective form of therapy, at least for recurrent melanoma with or without satellitosis or in-transit metastases, and also in cases of distant isolated metastases.

6. DISCUSSION

Viewed clinically, the results presented above confirm that heat can cause regression of tumors. The first question that arises is the

Table 10. Melanoma: Hyperthermic perfusion with chemotherapy

Case	Site	TNM classification	Time of hyperthermia	Antiblastic agent	Subsequent therapy	Distant results	Elapsed time
1	Foot	$T_3N_1M_O$	4 h 41.5-42.4°C	Glyceraldheide	Amputation 10 days later for vascular insufficiency	N.E.D.	7 years 6 months
2	Foot	$T_3N_1M_O$	2 h 50 min 42-42.7°C	Melphalan	Refused further treatment	D.O.D.	1 year 4 months
3	Leg	$T_3N_1M_O$	3 h 15 min 41.8-42.6°C	Melphalan	Refused further treatment	D.O.D.	2 years 1 month
4	Leg	$T_3N_1M_O$	3 h 45 min 42.3-42.6°C	Melphalan	Disarticulation for recurrence at another hospital	D.O.D.	3 years
5	Thigh	$T_ON_OM_1$	3 h 50 min 42°C	Melphalan	Excision of recurrence; systemic chemotherapy	D.O.D.	2 years 4 months
6	Leg	$T_ON_3M_1$	2 h 15 min 41.5-42.4°C	Melphalan	Excision of recurrence; systemic chemotherapy	D.O.D.	2 years 9 months

N.E.D. = No evidence of disease.
D.O.D. = Died of disease.

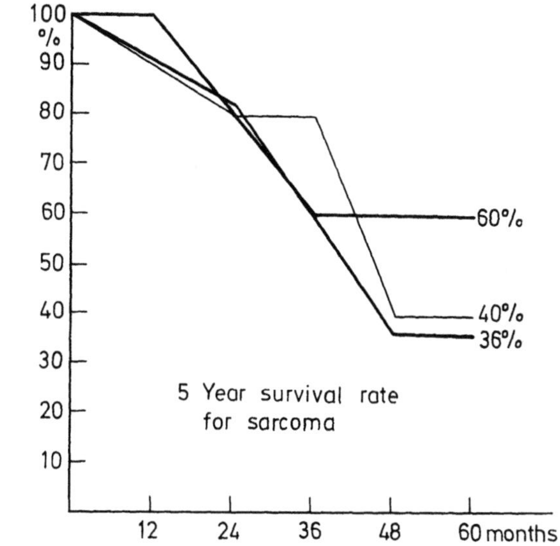

Fig. 5. (Text see p. 144)

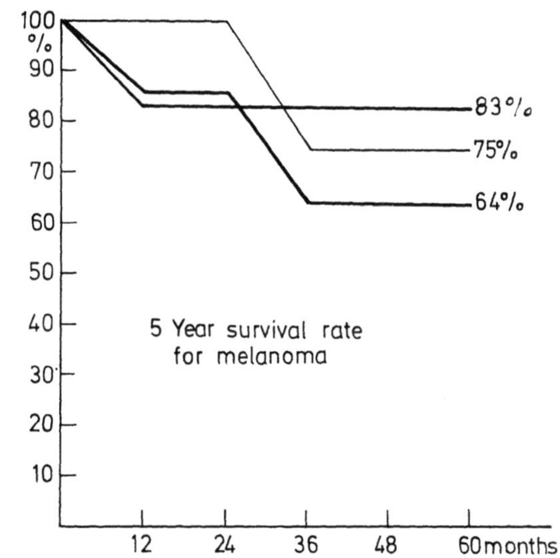

Fig. 6. (Text see pp. 146-148)

technique of applying local heating. Experimentally the results obtained by hyperthermic perfusion were very satisfactory. In fact, we have seen that the transplantability of IRE sarcoma or Yoshida sarcoma in rats is completely inhibited after 30-45 min of hyperthermic perfusion. This does not happen after the same period of time at the same temperature with water bath or diathermy. Appropriate here is CRILE's observation (6) concerning the injection of serotonin into S-180 tumors; when such an injection is performed before heating, only half the time of exposure to heat is required to destroy the tumor. The same role of a vasoactive drug may perhaps be played by the direct action of heat on the vessel walls, or indirectly through the acidosis which follows the microcirculatory modifications caused by hyperthermic perfusion. Furthermore it should be emphasized that diathermy, as a

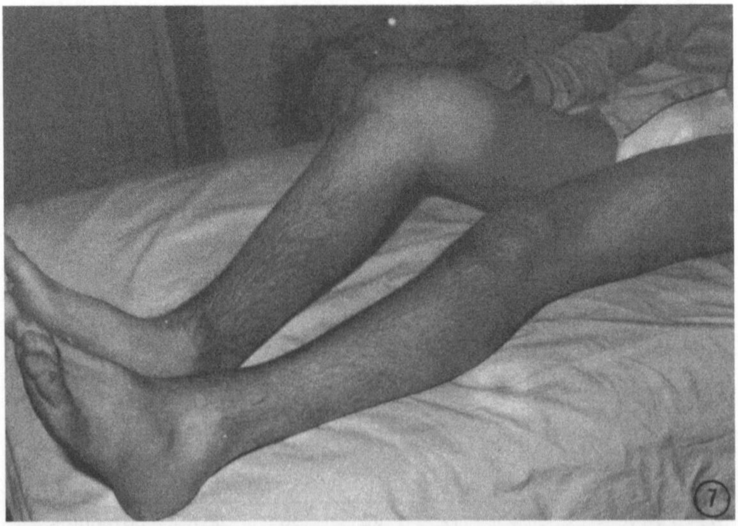

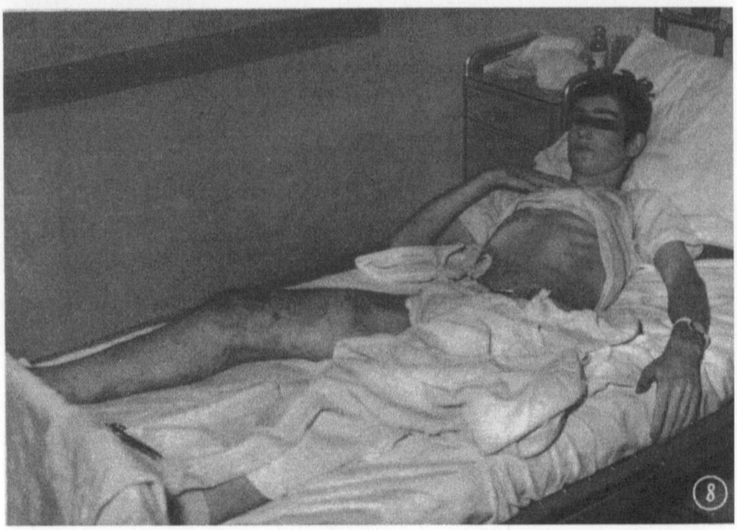

Fig. 7. V.Luigi. Osteosarcoma of right femur
Fig. 8. Four weeks after biopsy and hyperthermic perfusion. Patient was then amputated. No tumor was found at histologic control

method of heating, is not reliable because the temperature is not uniform throughout the tumor. This means that hyperthermic treatment cannot be effective unless all cancer cells can be uniformly maintained at high temperatures to obtain the destruction of the tumor. This is confirmed by DICKSON and ELLIS (10), who stated that there is a heating range below 41.5°C but above normal temperature in which metabolism is increased. Maintaining Yoshida tumors on the feet of rats at 40°C for 1 h, they observed an increase in both respiration and glycolysis of the tumor followed by enhanced dissemination of the cancer. The inadequate heating may explain also the recurrence of tumor. According to SKIPPER (32) the same inadequacy could enhance the dissemination of cancer. PALZER and HEIDELBERGER (26) state that cells recover from sublethal hyperthermic damage. According to these authors hyperthermic destruction is at least a two-step process: a potential lethal damage appears to be converted into lethal damage after a suitable time.

Repairing systems appear to be involved in the thermosensitivity of tumor cells (14). Conversely, at temperatures over 42°C, the host's tolerance must be taken into account, together with the decrease in selective sensitivity of malignant cells. It is our opinion that these problems are overcome by perfusion which seems to be a well-standardizing heating technique, capable of achieving and maintaining homogeneous temperatures within a controlled range, in order to obtain at least at the experimental level, total necrosis of the tumor with no viable cells after treatment. Also, in human limb tumors hyperthermic perfusion could be the method of choice if heat treatment has to be applied alone or in combination with chemotherapy. As for combined hyperthermia and radiation, we probably need more information concerning their interrelationships, as stated by SUIT and SCHWAYDER (33), before standardizing any appropriate clinically useful technique.

Legends to Figs. 9 to 12 on page 136.

Fig. 9. Osteosarcoma of lower end of femur in a 21-year-old man showing formation of osseous and osteoid tissue by tumor cells. Area of cellular pleomorphism with many multinucleated forms

Fig. 10. Same tissue as Fig. 9, showing an area of more abundant osseous and osteoid differentiation. Malignant osteoblasts tend to align around masses of osteoid tissue

Fig. 11. Same tumor 48 h after treatment. An area composed of degenerating cells, some with diluted blood sinusoids and spicules of calcified and necrotic bone

Fig. 12. Same tumor 1 month after treatment in amputated limb. Osseous trabeculae are thick, irregularly contoured, and set in relatively little supporting connective tissue containing hemosiderin deposits, scattered small dark-staining cells resembling lymphocytes, plasma cells, and eosinophils

Remarks. Osteosarcoma is very sensitive to high temperature perfusion. After treatment a variable quantity of necrotic bony trabeculae, separated by fibrovascular connective tissue, is present in the mass of the tumor. In collagenized and poorly osseous tissue areas one is likely to observe microscopic foci of edema, degenerating atypical cells, dilated blood sinusoids, and sometimes hemorrhage. Necrosis of bone occurs in irregular large or small patchy areas. There is seldom a clear line of demarcation between neoplastic and normal bone, both necrotic. Later one finds that the osseous trabeculae are thick, irregularly contoured, and set in relatively little supporting connective tissue. This is reduced to a very loose reticulum, fibrous-like containing hemosiderin deposits, plasma cells, and eosinophils

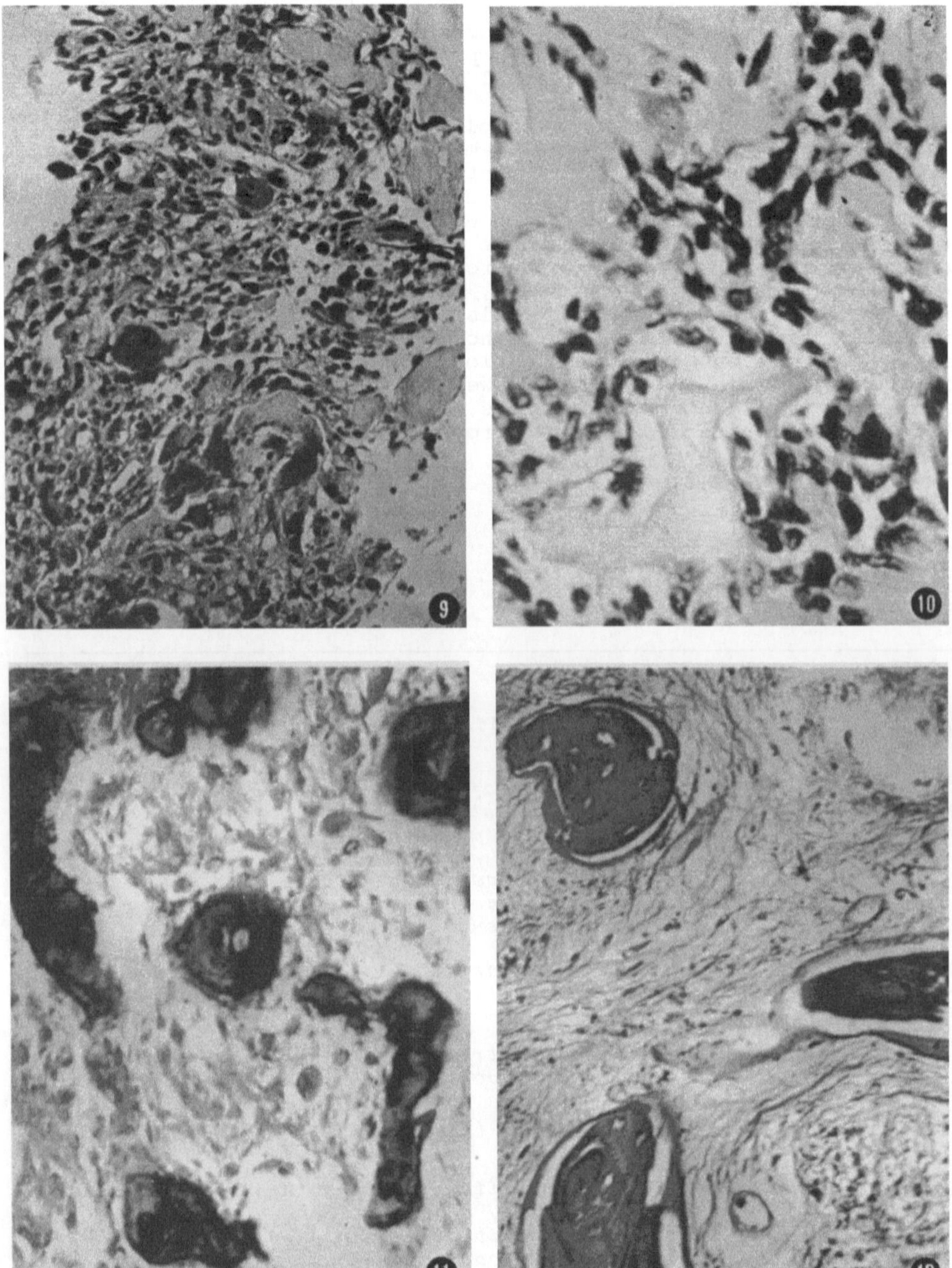

Figs. 9 to 12. (Legends see page 135)

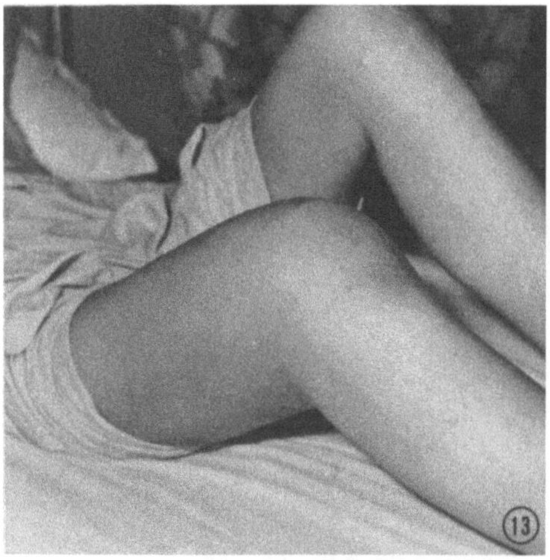

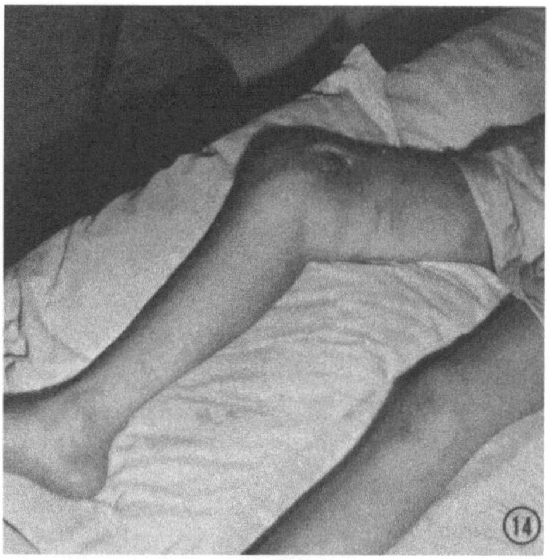

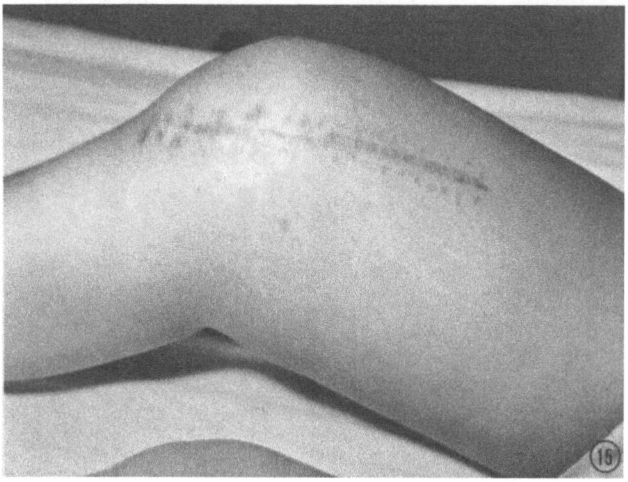

Fig. 13. P. Antonia. Chondrosarcoma of femur
Fig. 14. Four weeks after biopsy and hyperthermic perfusion
Fig. 15. Six months after excision of tumor

When applying hyperthermic perfusion one must take into account the possibility of local or general damage. This aspect of the problem cannot be ignored even if major complications caused by heat treatment actually disappear, as described above. A question still awaiting solution is how to maintain a perfused limb in good condition while achieving the maximum effect of heat on the tumor. It is probably an insoluble problem, because the action of heat apparently involves not only a direct destructive mechanism on cancer cells but also needs some reaction from the heated tissues. Perhaps, besides microcirculatory changes, this is to identified with an inflammatory reaction mediated through metabolic pathways. The inflammatory reaction of the perfused limb should not be avoided, but considered as one of the components of hyperthermic treatment (see photomicrographs). However, the condition of the perfused limb, with our current technique, gradually improve the first week until complete recovery.

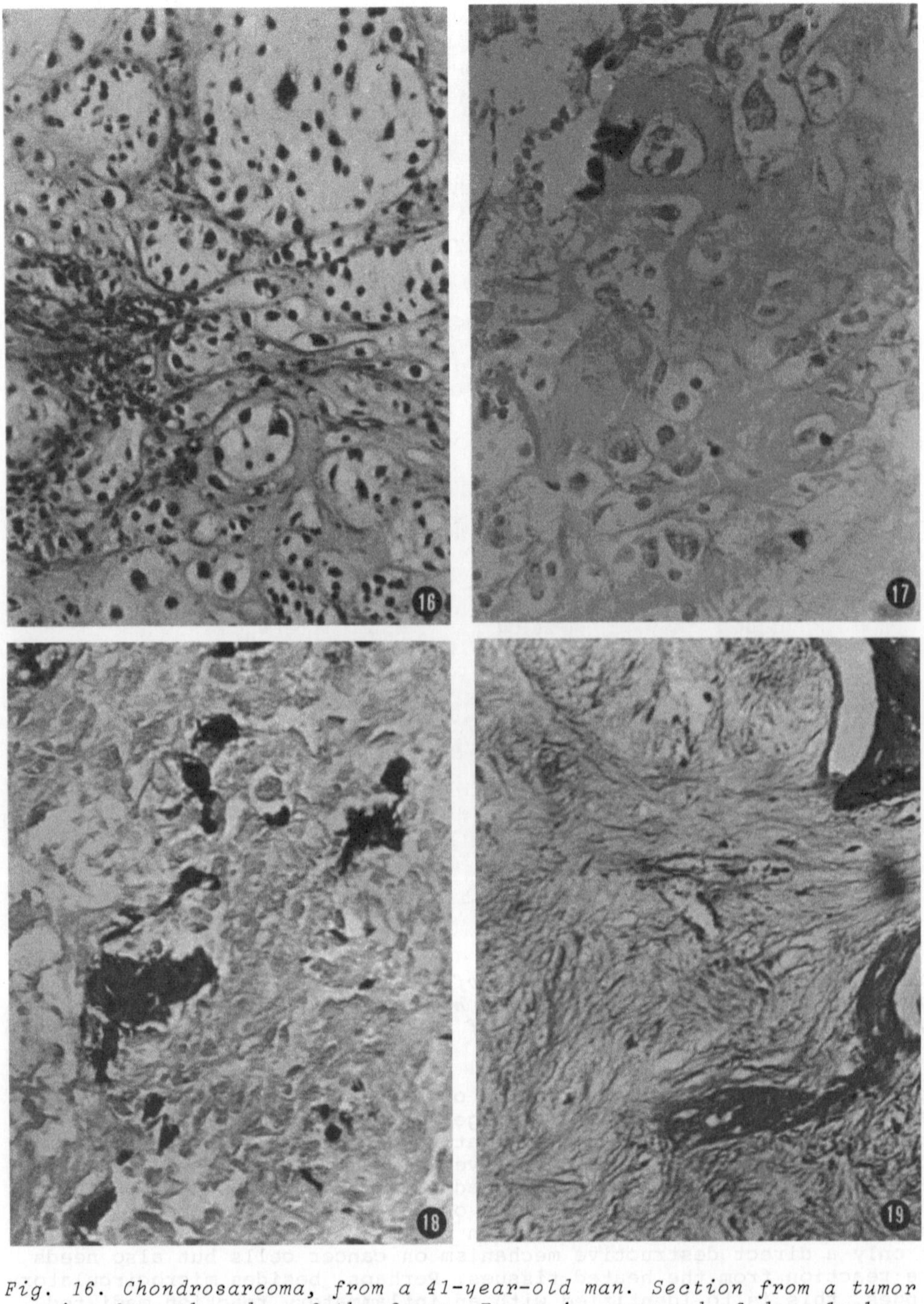

Fig. 16. Chondrosarcoma, from a 41-year-old man. Section from a tumor growing from the edge of the femur. Tumor is composed of densely basophilic cartilaginous islands lying in a clear hyaline matrix. There are areas of cytological pleomorphism; many lacunae contain cells in the process of division, some with multiple hyperchromatic nuclei. Other areas show a tendency for each cell to be surrounded by a separate mass of basophilic cartilage

Legends to Figs. 17-19 on page 138.

Fig. 17. Same tumor 48 h after perfusion. Extensive degeneration of tumor, which resembles an aseptic necrosis or an infarct. Neoplastic cells are swelling, embedded in a somewhat hyaline matrix, and shadow-like in appearance

Fig. 18. Same tumor 7 days after perfusion. Small and large areas of calcification are present with necrotic debris, partially collagenized, and with appearance of a dense and more or less homogeneous gel

Fig. 19. Same tumor 1 month after treatment in the amputated limb. A thin layer of poorly formed, still viable bone over the regions of the tumor is present. Dense fibrous tissue containing only scattered lymphocytes and plasma cells has replaced the mass of the tumor

Remarks. Chondrosarcoma varies greatly in its rate of growth and behavior. Therefore our evaluation of results obtaining with high temperature is very conservative. The histological picture after 7 days showed necrotic tissue with presence of focal areas of calcification. Furthermore, in the necrotic areas, there was evidence of collagenization, reabsorption of necrotic debris, and replacement of the latter by hyaline connective tissue

---

Legends to Figs. 20-23 on page 140.

Fig. 20. Fibrosarcoma of forearm in 39-year-old man. Tumor is highly cellular. Cells show considerable variation in size and shape and many of them appear typical. They are arranged in short parallel lines, in smooth curves, and in angular bends. Nuclei are elliptical and acicular, and exhibit many mitoses

Fig. 21. Same tumor 48 h after perfusion. Stroma is edematous and shows an inflammatory aspect with diffuse cellular infiltration composed of lymphocytes, plasma cells, and some histiocytes. Neoplastic cells are enlarged with wrinkled nuclei and vacuolized cytoplasm and are intermingled with inflammatory cells

Fig. 22. Same tumor 20 days after perfusion. Neoplastic tissue is replaced by a fibrous matric partially infiltrated by lymphocytes. On the right a few spindle-shaped cells (fibroblasts) are surrounded by numerous collagen bundles

Fig. 23. Same tumor 40 days after perfusion. The area occupied by the tumor is extensively fibrous in appearance

Remarks. The immediate effects of hyperthermia on neoplastic tissue of fibrosarcoma are the following: a) marked cellular enlargement accompanied by nuclear swelling, b) wrinkling of nuclei with breakage of nuclear membrane, c) vacuolization of cytoplasm or of nucleus or multinucleation and appearance of bizzare cell forms. The homogeneity and pallor of the enlarged nuclei are an especially valuable morphologic manifestation and occasionally one can observe a phenomenon of "nuclear explosion," i.e., extensive fragmentation of the nucleus. The stroma is edematous and has an inflammatory aspect with dilated blood vessels, newly formed capillaries, and extravasated erythrocytes. In addition, an inflammatory reaction may give the appearance of granulation tissue. There is perivascular as well as diffuse cellular infiltration composed of lymphocytes and fibroblasts. In late lesions, the histologic picture shows slight proliferation of spindle-

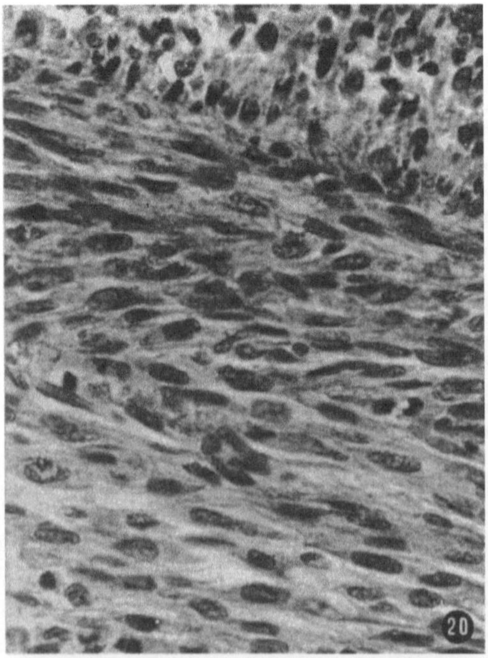

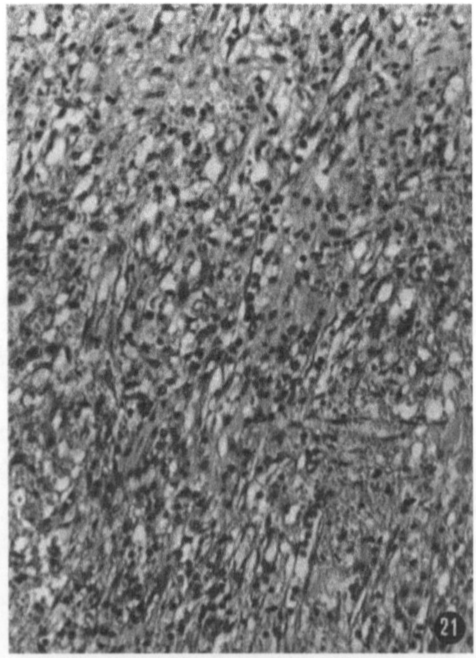

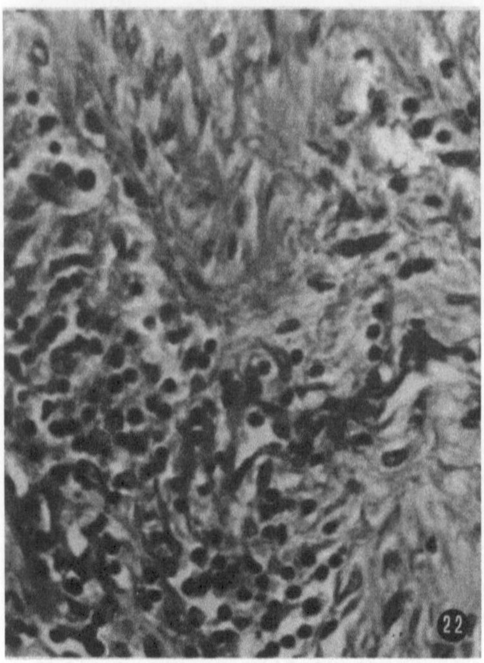

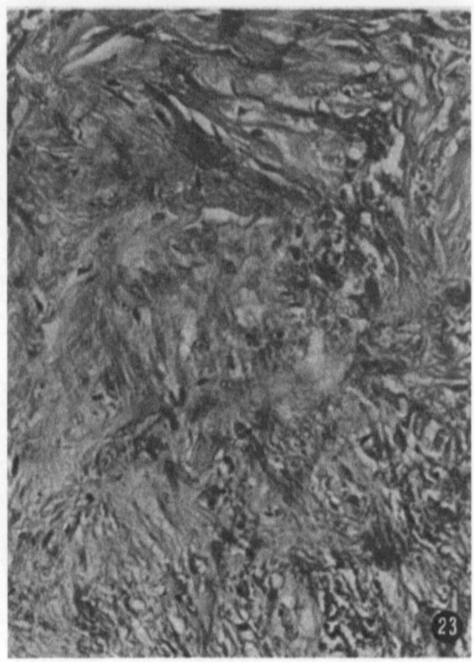

shaped cells that represent nonneoplastic young fibroblasts. They lie in strands that extend irregularly in all directions and are surrounded by numerous collagen bundles. Finally, in more favorable cases, the area previously occupied by the tumor is extensively fibrous in appearance

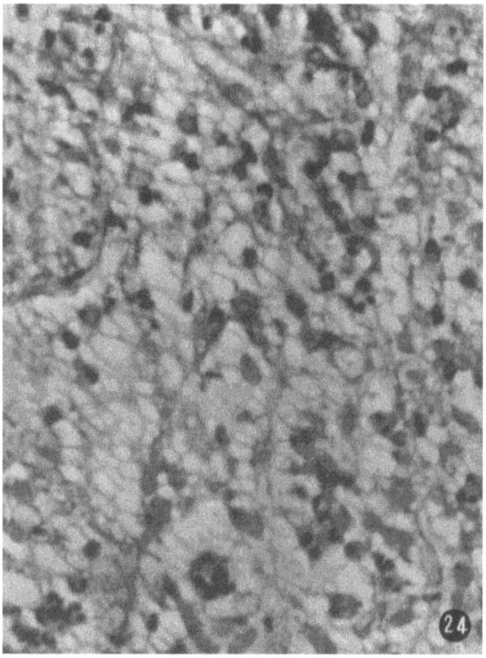

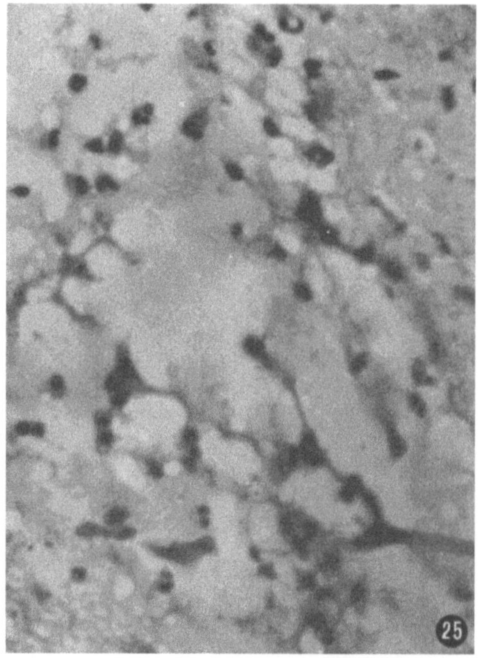

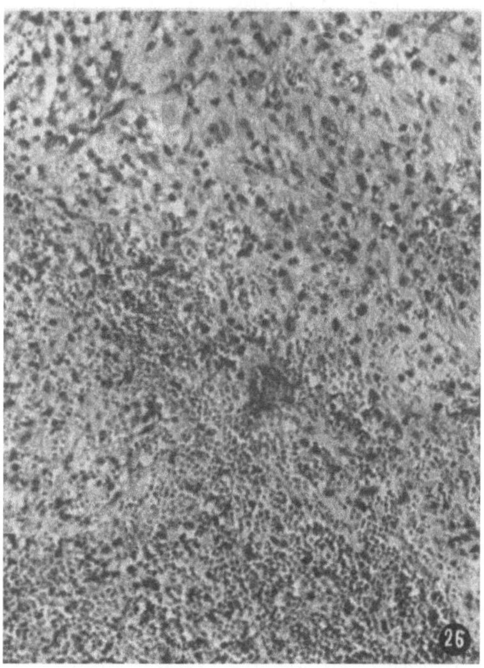

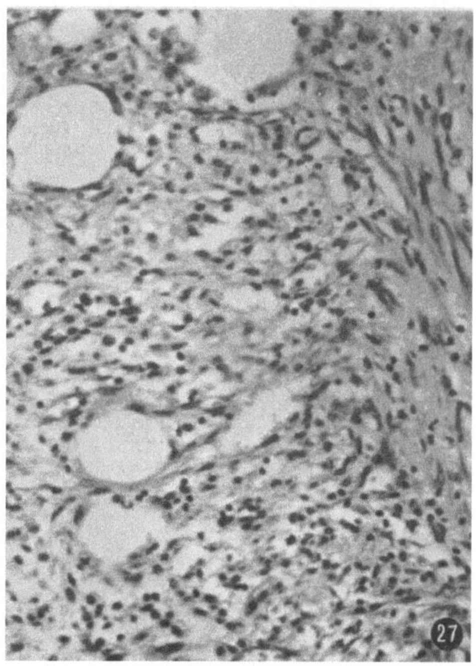

Fig. 24. Liposarcoma of thigh in 53-year-old woman developing in the intermuscular posterior planes. Tumor is composed largely of small embryonal spheroidal fat cells arranged diffusely; in some areas there is a mixture of adult and embryonal fat cells

Fig. 25. In this microscopic field one sees predominantly myxoid types of liposarcoma infiltrating the muscle. Spindle and stellate lipoblasts and more mature fat cells are set in a mucilaginous matrix

Legends to Fig. 26 and 27 on page 141.

Fig. 26. Same tumor 48 h after treatment. Neoplastic cells appear swollen and sometimes granular. There is a cellular infiltrate composed of lymphocytes, histiocytes, fibroblasts, and extravasated erythrocytes

Fig. 27. Same tumor removed 25 days after perfusion. There is a cellular infiltration, perivascular as well as diffuse, composed of lymphocytes, plasma cells, and some histiocytes and fibroblasts. The appearance is of granulation tissue. The many cavities, large and small, ovoid or round, that give the section a "Swiss cheese" appearance, are lipid spaces

Remarks. After exposure to high temperatures poorly defined areas of necrobiosis of collagen and adipose tissue are present. Collagen bundles appear split, amorphous, and intermingled with fibrin deposits. There is, at the margin of the areas of necrobiosis, a cellular infiltrate composed of lymphocytes, histiocytes, fibroblasts, and occasional groups of epithelioid cells. Scattered foreign-body giant cells are sometimes present. Blood vessels exhibit thickening of walls with proliferation of endothelial cells. This process may lead to partial and, occasionally, even to complete occlusion of the lumen. Neoplastic cells appear swollen, myxomatous, multivacuolated, and sometimes granulated. Destruction of degenerate lipoblasts from still active cells may be very difficult

---

On the basis of our clinical experience a different way of proceeding is necessary for the treatment of osteogenic sarcoma. Here the best results are obtained with hyperthermic perfusion followed some time later by amputation. In patients treated for melanoma, the best results were obtained with hyperthermic perfusion alone or followed by excision some weeks later. On the other hand, when excision was performed immediately after hyperthermic perfusion, all patients died. Even though these data should be confirmed by a more extensive list of cases, they lead us to conclude that the increased immunogenicity of tumor cells plays a role in the complex mechanism of the action of heat. Such a role can be more or less closely linked to type, size, and localization of the tumor. Osteogenic sarcoma is probably affected by heat and inflammatory reaction to a lesser extent than other limb tumors on account of its particular vascularity and localization. Amputation is therefore necessary, but only after a period of time during which the presence of the heated tumor is useful as an antigenic source for immunization of the host against metastases. In melanoma the direct and indirect action of heat may be sufficient to prevent metastases. In soft tissue sarcoma the results do not show significant variations, as far as the time at which the excision was performed is concerned. This is presumably due to the incompleteness of the immediate excisions that were carried out to avoid major renal damage by reabsorption of the tumor. In these respects our results agree with CRILE (6), MATHE (21), GREENSTEIN (15), WESTERMARK (38), and others.

Another point which should be discussed concerns the possibility of combining antiblastic drugs with hyperthermic perfusion. Our experience here is limited to the treatment of melanoma. In tumors other than melanoma we have performed antiblastic hyperthermic perfusions only recently, and therefore cannot express an opinion. As described above, we treated a group of patients affected by melanoma with two different techniques of combined hyperthermia and chemotherapy. There is no doubt that introducing the drug during the last hour of hyperthermic perfusion is contraindicated. In fact in these cases we found much more local damage without appreciable improvement in distant results.

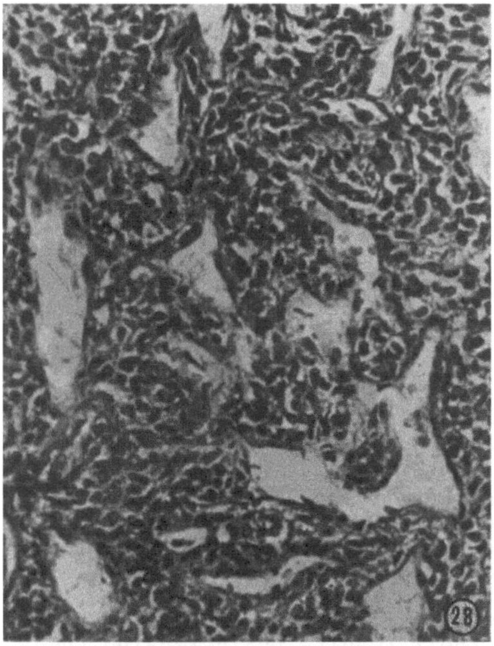
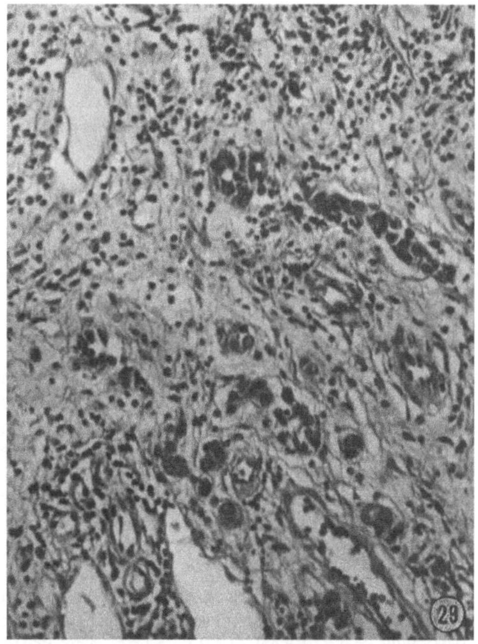
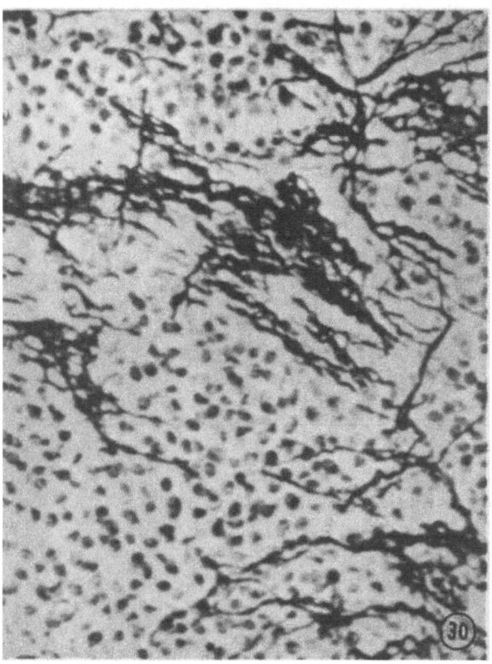
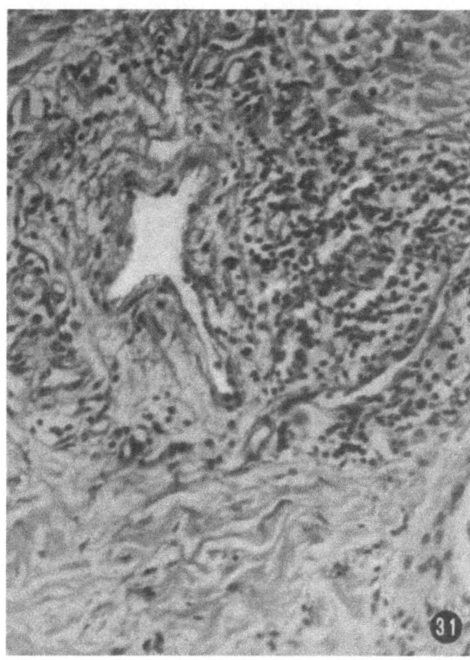

Fig. 28. Hemangiopericytoma of leg in a 30-year-old man. Relapsed 2 months after local excision. Tumor is characterized by a rich vasculature consisting of endothelial channels surrounded by round, ovoid, or spindle-shaped cells with prominent vescicular nuclei

Fig. 29. Same tumor 48 h after perfusion, showing degeneration and swelling of collagen in walls of blood vessels. Vascular lumina are narrowed and normal vessels appear side-by-side

Legends to Fig. 30 and 31 on page 143.

Fig. 30. Reticular pattern of previous figure showing breakage of fibrils. One does not recognize distinctive reticulin pattern running around blood vessels

Fig. 31. Tumor area 1 month after perfusion. Extensive fibrosis and a starry vessel surrounded by many lymphocytes
Remarks. Tumor after perfusion shows degeneration and swelling of collagen and smooth muscle in walls of blood vessels. Swollen endothelial cells narrow the lumina which appear side-by-side. There is a marked interstitial edema with sparse infiltration by extravasated erythrocytes. Later, extensive fibrosis occurred, containing cells resembling lymphocytes

---

The only patient free of disease was treated with glyceraldehyde. The condition of his limb, which had to be amputated, and his postoperative general condition made us resolve never to use this drug again. The other technique of antiblastic hyperthermic perfusion which we carried out is the one introduced by STEHLIN (31). In these cases our evaluation does not take the distant results into consideration because not enough time has elapsed since the patients underwent antiblastic hyperthemic perfusion. At present the immediate results are satisfactory, but they seem to hold for a limited period only. The synergistic action on immunoresponse of antiblastic and high temperature simultaneously administered in perfusion was observed by STEHLIN (31). KREMENTZ and RYAN (19) also improved their results by increasing the temperature in the perfusion circuit. The long-term results of these authors are impressive but we have not been able to reproduce them in our material. In order to assess our results we evaluated statistically by actuarial methods the 5-year survival of the patients treated (Fig. 5). The 5-year survival rate for soft tissue sarcomas (about 40%) is lower than in other tumors probably because the excision of the tumor is currently performed immediately after perfusion in order to avoid reabsorption phenomena, thus decreasing the immunoresponse evoked by the heated tumor. For all osteogenic sarcomas we obtained a 5-year survival rate of 36%, but it must be emphasized that the survival rate increases to about 60% if we consider only the group of osteogenic sarcomas treated by hyperthermic perfusion systematically followed by amputation 4 weeks later. This is a very high 5-year survival rate for this kind of tumor, and it indicates that hyperthermic perfusion followed by amputation could be the most effective way of

---

Legends to Figs. 32 - 35 on page 145.

Fig. 32. Rhabdomyosarcoma of quadriceps occurring in 49-year-old-man. Tumefaction has been present for 2 months and has grown to the size of a small melon. An area of cellular pleomorphism in which are many multinucleated giant cells and spindle-shaped cells randomly arranged without pattern
Fig. 33. Same tumor 48 h after perfusion, showing marked interstitial edema with sparse infiltration by extravasated erythrocyates and some giant cells
Fig. 34. Another area showing cytologic evidence of hyperthermal injury. Main characteristics are: 1) nuclear swelling; 2) rupture of nuclear membranes; 3) granular or homogeneous appearance of cytoplasm
Fig. 35. Same tumor 1 month after treatment in amputated limb. Fibrous tissue, rich in vessels and infiltrated by lymphocytes and plasma cells has replaced tumor tissue

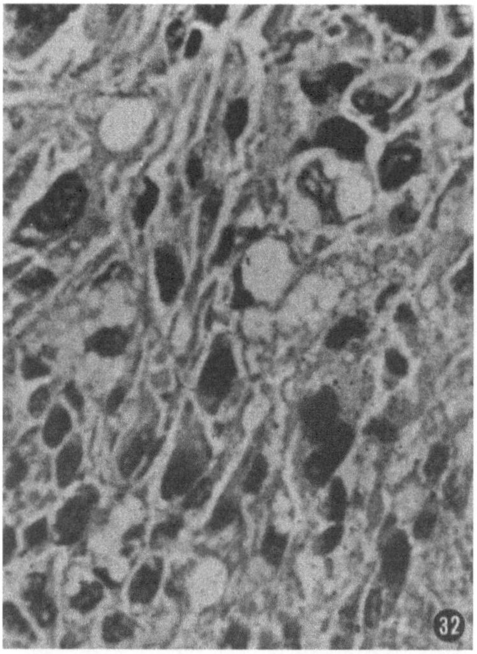

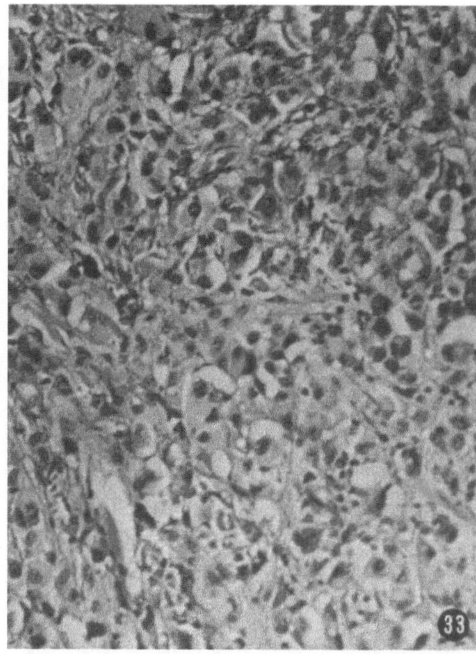

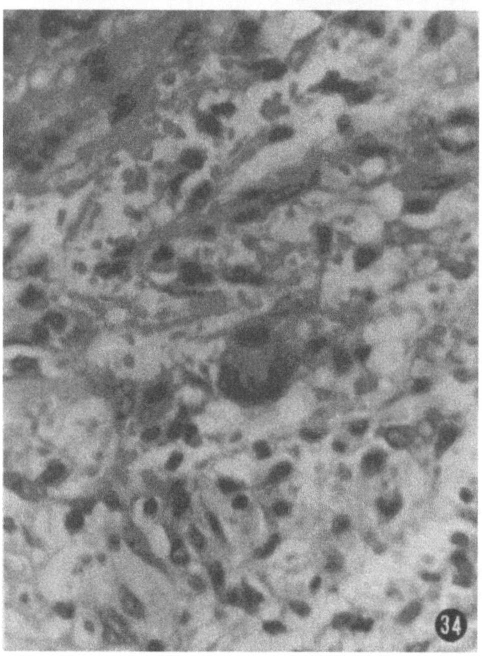

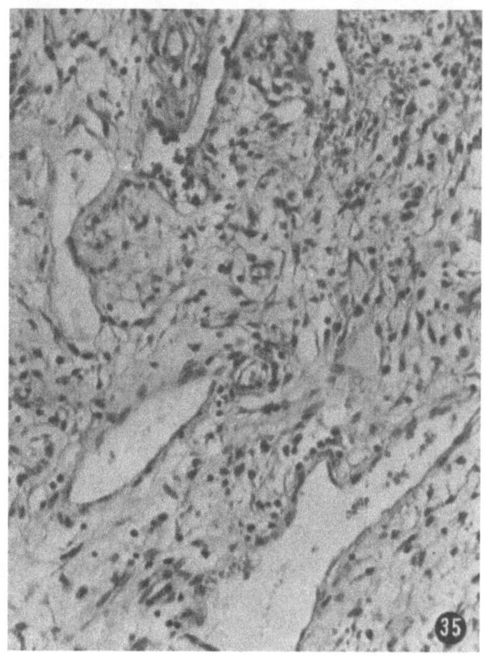

Remarks: Rhabdomyosarcomatous neoplasms that arise in relation to skeletal muscles of adults after perfusion show cytological alterations such as nuclear swelling, rupture of nuclear membranes, and finally pyknosis. Cytoplasm of thermally injured cells becomes at first granular and later homogenously coagulated. Later, fibrosis with sparse cellularity occupies tumoral area

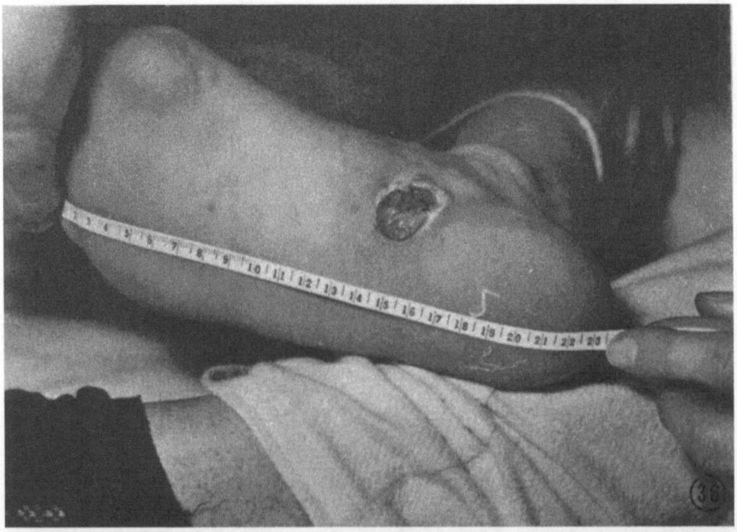

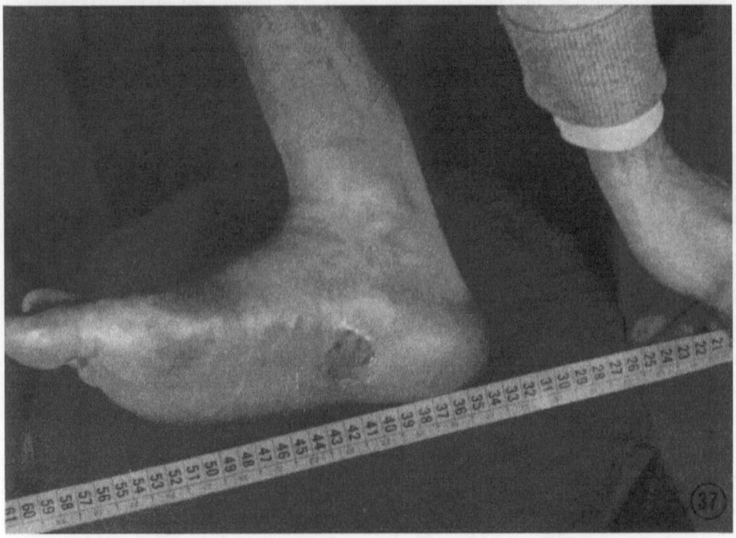

Fig. 36. L. Aurelio. Melanoma of foot
Fig. 37. Same as Fig. 36. 6 months after hyperthermic perfusion

treating human osteogenic sarcoma. However, it is a fact that the results shown, which we hope will be confirmed by more extensive trials, have not been achieved with any other kind of combined therapy that has been proposed for the treatment of osteogenic sarcoma to date.

The same statistical evaluation of 5-year survivals was made for melanoma (Fig. 6). The best results were obtained with hyperthermic perfusion alone and with hyperthermic perfusión followed by excision 4 weeks later. The latter procedure gave a 5-year survival rate of about 75%, less than in hyperthermic perfusion alone in which the 5-year survival is 83%. On the other hand, if we evaluate also

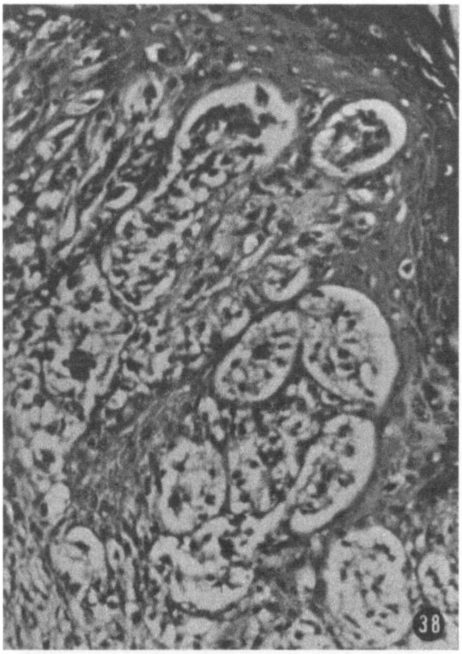

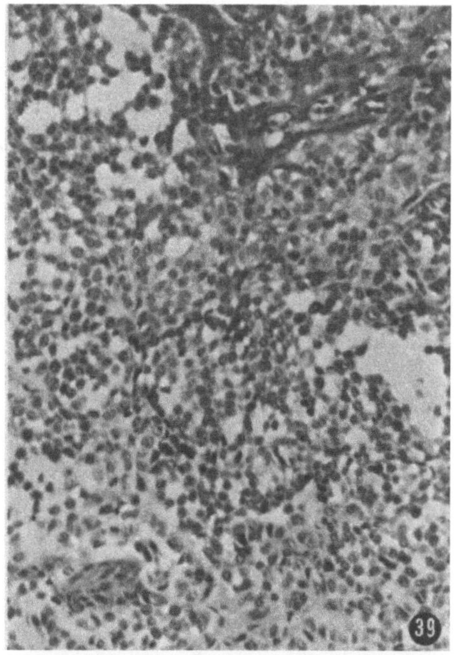

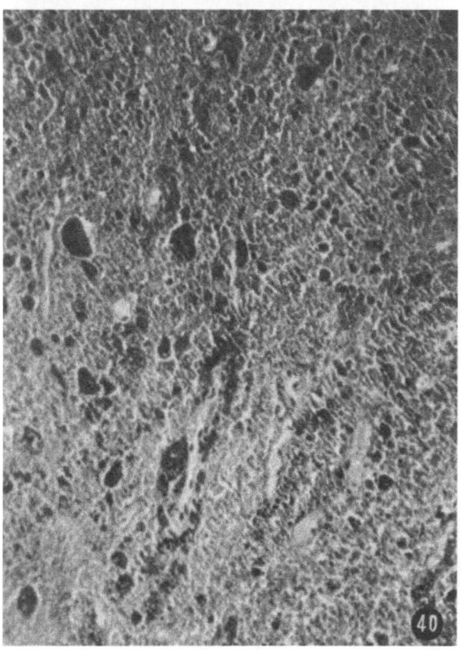

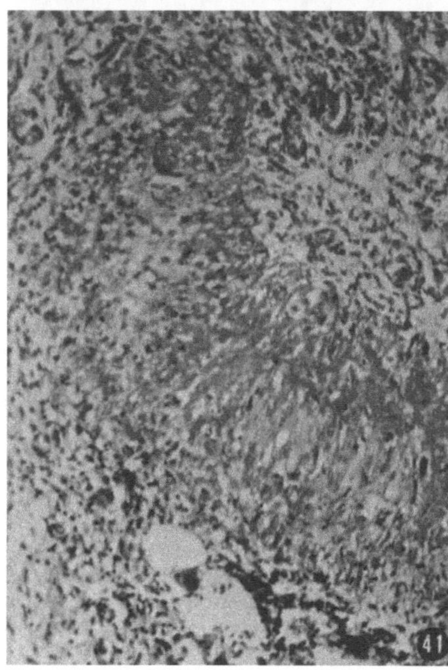

Fig. 38. Melanoma of sole of foot in 51-year-old man. Edge of lesion shows migration of pigmented and clear melanocytes from epidermo-dermal junction into corium

Fig. 39. Same tumor. Section from deep part of lesion shows that tumor cells are largely of cuboid type and lie in alveolar formations

Legends to Fig. 40 and 41 on page 147.

Fig. 40. Area of tumor 48 h after perfusion, showing an extensive necrosis which is destroying a large quantity of tumor tissue, so that its cellular pattern is unrecognizable. Sparse melanin is present

Fig. 41. Same tumor 48 h after perfusion. An area of fibrinoid degeneration is surrounded by an area of edematous collagen infiltrated by inflammatory cells

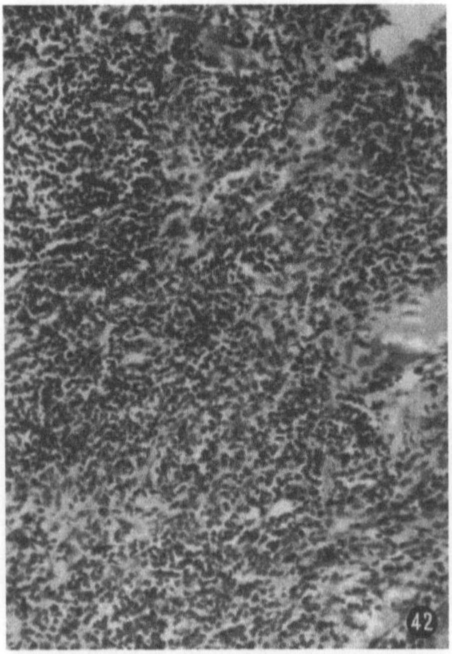

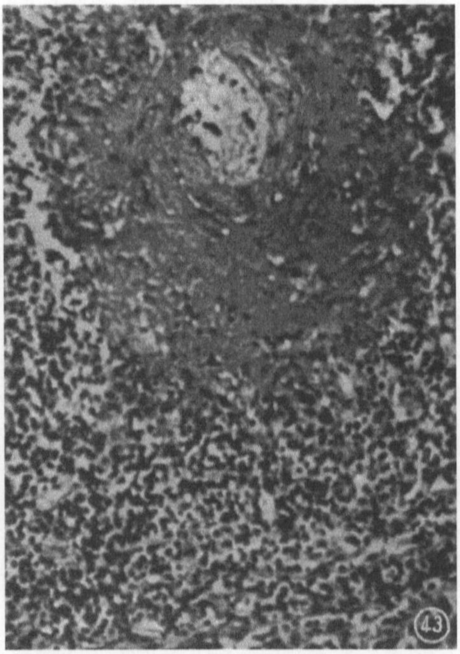

Fig. 42. Same tumor 5 days after perfusion. Note large patches composed almost entirely of lymphocytes. In addition, a few histiocytes and plasma cells are present. Note resemblance with halo nevus

Fig. 43. Same tumor 10 days after perfusion. The vascular walls of an artery show fibrinoid necrosis. Lymphocytes and plasma cells surround the vessel

the total group of melanomas, the 5-year survival rate decreases to about 60%. This confirms our previous observations, the the presence of the treated tumor is really important, probably because it acts as an antigenic source to immunize the host against recurrences and metastases. Indeed, in our opinion, hyperthermic perfusion alone is more reliable for the treatment of melanoma.

It is a fact that heating can be important in cancer therapy, even though a general standardization of its application is still far off, eccept for some limb tumors. Our present knowledge and the prospects concerning, on the one hand, the various treatments associated with hyperthermia and, on the other, the physiopathology and therapeutic action of whole-body treatment, appear to be promising.

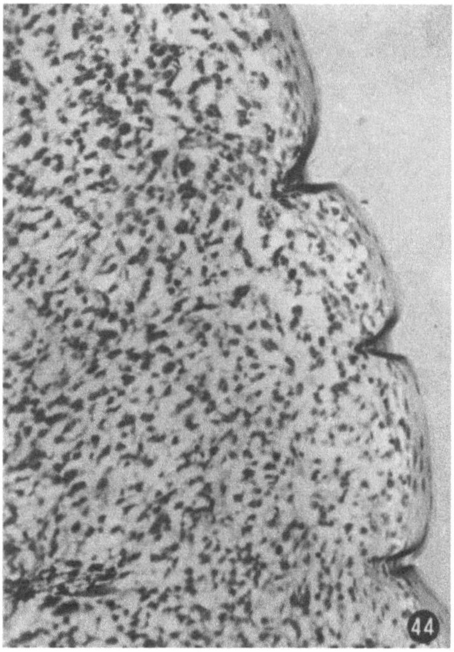

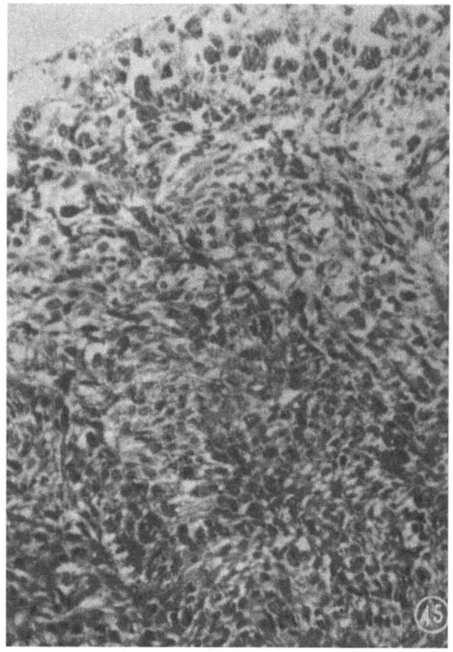

Fig. 44. Melanoma of forearm in 47-year-old man. Tumor cells are fusiform and lie in irregular strands

Fig. 45. Deep part of same tumor. Spindle cells occur haphazardly and in alveolar groupings

---

Legends to Figs. 46-49 on page 150.

Fig. 46. Same tumor 48 h after perfusion. Necrotic changes have taken place in this area. One recognizes only scattered altered tumor cells, sparse melanin and well-preserved hair follicle

Fig. 47. An area of tumor 7 days after perfusion. Many necrotic fat cells showing a cloudy appearance can be easily recognized. Some resemble balloon cells

Fig. 48. One month after perfusion. Tumor area is replaced by granulation tissue with giant cells. Some are Langhans', others foreign-body type

Fig. 49. Another area of same tumor. Bizarre giant cells are present, perhaps originating from muscle fibers

Remarks. The main modifications and most striking histologic features are areas of necrosis which destroy a large amount of tumor tissue or else so distort its cellular pattern that identification is impossible. A striking feature of a treated melanoma is often the presence of sparse melanin, extruded by damaged melanocytes, and collected by accompanying chromatophores

Perfused neoplastic tissue in hours immediately subsequent to perfusion shows, in addition to necrosis, a dense polymorphous infiltrate, formed by neutrophils, histiocytes, lymphocytes, and plasma cells. Frequently the nuclei of some neutrophils are fragmented, thus forming nuclear dust, especially in the vicinity of the capillaries. Many of these are dilated and show strongly eosinophilic fibrinoid material within and around their walls

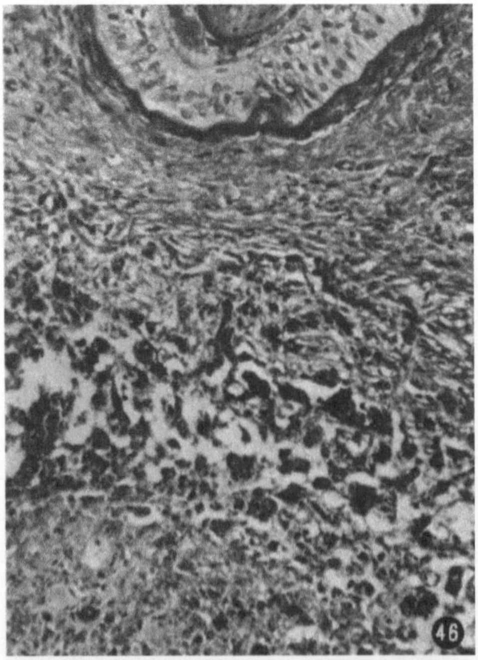

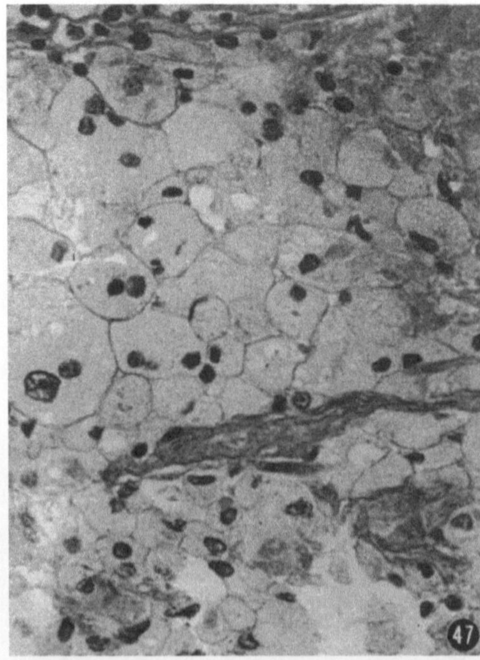

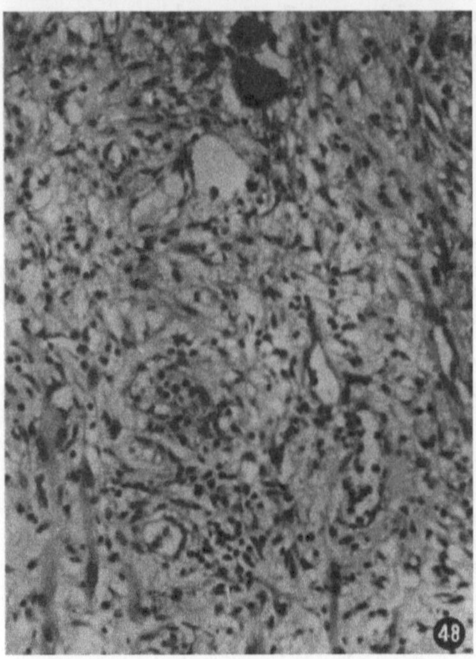

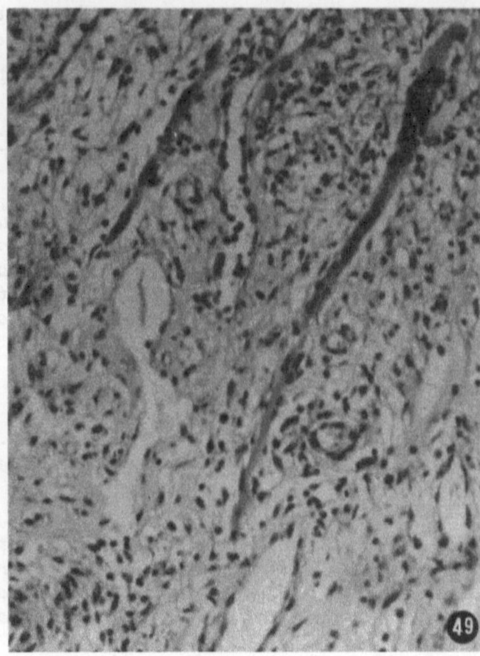

Many extravasated red cells are present. At the last stage the capillaries often show proliferation of their endothelium and the infiltrate is broken up into patches of various sizes by strands of collagen. Number of neutrophils is greatly reduced, and plasma cells, lymphocytes, and fibroblasta predominate.

It is noteworthy that in some cases the vascular walls of medium-sized and small arteries show focal lesions similar to panarteritis nodosa. In fact there is fibrinoid necrosis of the media, destruction of the internal elastic lamina, and leukocytic infiltration into the walls of vessels. In adventitia, too, there is fibroblastic proliferation, merging with intimal growth

At a distance, granulating areas with giant cells, like Langhans' or bizarre in shape, replace the tumor tissue

REFERENCES

1. BUSCH, W.: Über den Einfluss, welchen heftigere Erysipeln zuweilen auf organisierte Neubildungen ausüben. Verhandl. Naturh. Preuss. Rhein. Westphal. 23, 28-30 (1866)
2. BURNS, P.: Die Heilwirkung des Erysipels auf Geschwülste. Beitr. klin, Chir. 3, 443-446 (1887)
3. COLEY, W.R.: The treatment of malignant tumours by repeated inoculation of erysipelas. With a report of ten original cases. Amer. J. Sci. 105, 487-511 (1893)
4. CAVALIERE, R., CIOCATTO, E.C., GIOVANELLA, B.C., HEIDELBERGER, C., JOHNSON, R.O., MARGOTTINI, M., MONDOVI, B., MORICCA, G., ROSSI-FANELLI, A.: Selective heat sensitivity of cancer cells. Cancer (Philad.) 20, 1351-1381 (1967)
5. CRILE, G.: The effects of heat and radiation on cancer implanted on the feet of mice. Cancer Res. 23, 372-380 (1963)
6. CRILE, G., Jr.: Selective destruction of cancer after exposure to heat. Ann. Surg. 156, 404-407 (1962)
7. CRILE, G., Jr.: Heat as an adjunct to the treatment of cancer. Experimental studies. Cleveland Clin. Quart. 28, 75-89 (1961)
8. CRILE, G.: Rationale of simple mastectomy without radiation for clinical stage 1 cancer. Surg. Gynec. Obstet. 120, 975-982 (1965)
9. CREECH, D., KREMENTZ, E.T.: Cancer Chemotherapy by perfusion. Advanc. Cancer Res. 6, 111-147 (1961)
10. DICKSON, J.A., ELLIS, H.A.: Stimulation of tumor cell dissemination by raised temperature in rats with transplanted Yoshida tumor. Nature (Lond.) 248, 354-358 (1974)
11. DICKSON, J.A., SUZANGAR, M.: "In Vitro in Vivo" studies of the susceptibility of the Yoshida solid sarcoma to drugs and hyperthermia (42°C). In press
12. EHERSON, H.C., COLE, W.E.: Spontaneous regression of cancer. Preliminary report. Ann. Surg. 144, 366-383 (1956)
13. FOWLER, G.A.: Effects of acute concurrent infection on cancer in man (Series A). Pyrogenic infection in inoperable cancer, 152 cases. Unpublished report
14. GIOVANELLA, B.C., MORGAN, C.A., STEHLIN, J.S., WILLIAMS, L.J.: Selective lethal effect of supranormal temperatures on mouse sarcoma cells. Cancer Res. 33, 2568-2578 (1973)
15. GREENSTEIN, J.P.: In: Biochemistry of Cancer. New York: Academic Press. (1954)
16. GOLDENBERG, D.M., LANGUER, M.Z.: Direct and abscopal antitumor action of local hyperthermia. Z. Naturforsch. B., 26, 359-361 (1971)
17. JOHNSON, H.T.: The action of short radio waves on tissues. III. A comparison of the thermal sensitivities of transpantable tumors in vivo and in vitro. Amer. J. Cancer 38, 533-550 (1940)
18. KIRSCH, R., SCHMIDT, D.: Erste Experimentelle und Klinische Erfahrungen mit den Gauzkörper - Extrem - Hypertermic. In: Aktuelle Probleme aus dem Gebiet der Cancerologie. Doerr, W.,

Lind, F., und Wagner, G., (eds.). Berlin-Heidelberg-New York: Springer-Verlag. 1966, pp. 53-70

19. KREMENTZ, E.T., RYAN, F.: Chemotherapy of melanoma of the extremities by perfusion: fourteen years of clinical experience Ann. Surg. 175, 900-917 (1972)
20. KVETINA, J., GUAITANI, A.: A versatile method for the in vitro perfusion of isolated organ of rats and mice with particular reference to liver. Pharmacology 2, 65-81 (1969)
21. MATHE, G.: Active immunotherapy. Advanc. Cancer Res. 14, 1-36 (1971)
22. NAUTS, H.C., FOWLER, G.A., BOGATKO, F.H.: A review of the influence of bacterial infections and bacterial products (Coley's toxin) on malignant tumours in man. Acta med. scand. Suppl. 267, 103 (1953)
23. OVERGAARD, K., OVERGAARD, J.: Investigations on the possibility of a thermic tumour therapy. II. Action of combined heat-roentgen treatment on a transplanted mouse mammary carcinoma. Europ. J. Cancer, 8, 573-576 (1972)
24. PERCY, J.F.: Heat in the treatment of carcinomas of the uterus. Surg. Gynec. Obstet. 22, 77-79 (1916)
25. PETTIGREW, R.T., GALT, J.M., LUDGATE, C.M., SMITH, A.N.: Clinical effects of whole body hyperthermia in advanced malignancy. Brit. med. J. 4, 679-682 (1974)
26. PALZER, R.J., HEIDELBERGER, C.: Studies on the quantitative biology of hyperthermic killing. Cancer Res. 33, 415-421 (1973)
27. ROHDENBURG, G.L.: Fluctuations in the growth of malignant tumours in man, with special reference to spontaneous recession. J. Cancer Res. 3, 193-225 (1918)
28. SELAWRY, O.S., CARLSON, J.C., MOORE, G.E.: Tumour response to ionizing rays at elevated temperatures. Amer. J. Roentgenol. 80, 82-89 (1958)
29. STEHLIN, J.S., CLARK, R.L., VICKERS, W.E., MONGES, A.: Perfusion for malignant melanoma of the extremities. Amer. J. Surg. 105, 607-614 (1963)
30. SHINGLETON, W.W., BRYAN, F.A., O'QUINN, W.L., KRUEGER, L.C.: Selective heating and cooling of tissues in cancer chemotherapy. Ann. Surg. 156, 408-416 (1962)
31. STEHLIN, J.S.: Hyperthermic perfusion with chemotherapy for cancer of the extremities. Surg. Gynec. Obstet. 129, 305-308 (1969)
32. SKIPPER, H.: In the proliferation and spread of neoplastic cells, twenty first annual symposium on fundamental cancer research, University of Texas, M.C. Anderson Hospital and Tumor Inst. Baltimore: Williams and Williams, 1968, p. 213
33. SUIT, H.D., SCHWAYDER, H.: Hyperthermia: Potential as an antitumor agent. Cancer (Philad.) 34, 122-129 (1974)
34. SUZUKI, K.: Application of heat to cancer chemotherapy. Experimental studies. Nagoya J. Med. 30, 1-21 (1967)
35. VIDAL, E.: Travaux de la Deuxième Conférence International pour l'Etude du Cancer. Paris, 1916, p. 160
36. WARREN, S.L.: Preliminary study of the effect of artificial fever upon hopeless tumour cases. Amer. J. Roentgenol. 33, 75-87 (1935)
37. WESTERMARK, F.: Über die Behandlung des ulzerierenden Cervix Karzinomes mittels konstanter Wärme. Zbl. Gynäk. 1336-1339 (1927)
38. WESTERMARK, H: The effect of heat upon rat tumor. Scand. Arch. Physiol. 52, 257-322 (1927)
39. WOODHALL, B., PICKGRILL, K.L., GEORIADE, N.G., DUKES, H.T.: Effect of hyperthermia upon cancer chemotherapy. Application of external cancer of head and face structures. Ann. Surg. 151, 750-759 (1960)

Chapter 6
Whole-Body Hyperthermia. A Systemic Treatment for Disseminated Cancer

R. T. PETTIGREW and C. M. LUGATE

Regression of a disseminated tumour following inter-current infection and prolonged high fever was first reported by BUSCH over 100 years ago. In 1893 COLEY confirmed this observation and in a series of 38 patients with extensive cancer, who developed erysipelas and high fever, 12 had complete regression. COLEY then tried to reproduce this effect by injecting a mixture of toxins from the organisms responsible for erysipelas into patients with advanced malignancy, effecting some remarkable regressions. The use of toxins was unsatisfactory as a considerable time was taken to reach a maximum temperature of 40°C, and this temperature, in a toxic patient with disseminated disease, was difficult to maintain for long treatment intervals.

An external, but controllable heat source was required which would conserve the patient's own energy.

1. METHODS OF INDUCING HYPERTHERMIA

Whole body hyperthermia may be induced by insulating the body to prevent heat loss and either increasing the normal rate of heat production or by applying an external heat source. At temperatures over 40°C there is a collapse of the normal thermoregulatory system of the body, and it is then necessary to provide an external temperature control system.

The injection of toxins, such as Coley's fluid, or hormones such as thyroxin, produces a slow rise to 40°C. The effects are unpredictable, especially when given to seriously ill patients, and there may be a stimulatory effect on the spread and growth of metastases at temperatures below 40°C (3).

Of the external sources of heat, microwaves and ultrasonics are the most difficult to control and cannot, at present, be used for whole body treatment. External sources otherwise rely on skin conduction. For efficient heat exchange the maximum area of heating is important as a skin temperature of over 45°C for any period of time causes local burning. Infra-red heating lamps exposed to 50% of the available area are not as efficient as an immersion technique.

There are three immersion techniques presently in use:

A. Water Baths

The immersion of the patient in hot water requires a complicated system for temperature control. As the patient is unable to evaporate sweat, due to the surrounding humidity, a constant flow technique is required whereby the temperature of the water in the bath can be changed quickly. It also must circulate, to prevent conduction difficulties. Water has the advantage that, because of buoyancy, there are few pressure areas. The patient is also exposed and there is easy access during treatment.

B. Hot Air Boxes

Air heating involves enclosing the patient in a constant stream of heated air. The patient must be insulated to prevent the evaporation of sweat, which otherwise requires excessive skin temperatures to be reached for efficient heating. This method usually requires a closed system and the accessibility of the patient is therefore reduced.

C. Wax Baths

Low melting point paraffin wax (M.O. 43-46°C), has the advantage of offering a large thermal resevoir, a low thermal gradient, and allows the free evaporation of sweat from any exposed area, enabling simple control of body temperature. At 50°C wax adjacent to the patient solidifies, insulating the skin from the higher temperature molten wax in the bath. As the wax solidifies it gives out latent heat, so keeping the temperature of remaining wax high. There is also some buoyancy afforded by the wax. Once a required temperature has been reached the molten wax can be removed leaving a thin pellicle which acts as an excellent insultor against heat loss from the body.

It is this method and the results achieved using this technique that are described in this chapter.

2. ANAESTHESIA DURING HYPERTHERMIA

At body temperatures above 41°C patients become restless and anxious. Increasing muscular activity makes external thermo-regulation more difficult. Light narcosis is therefore usually required for any lenth of treatment. At temperatures over 40°C there is a rapid increase in the quantitiy of anaesthetic agents required, although there is a decrease in liver metabolism (4).

Initially, short-acting barbiturates were given intermittently, but at the end of the treatment session, there was a large quantity of drug circulating, causing prolonged anaesthesia. The use of an epidural anaesthetic decreases the need for systemic anaesthetic agents. By causing peripheral vasodilation, the rate of heat absorption from the skin is increased and, by sympathetic block, the heart rate is controlled. Low dose halothene has also been used to good effect during hyperthermia (9).

3. METHOD

The patient is premedicated with promethazine hydrochloride, induced with thiopentone sodium, and curarized. An epidural catheter is introduced at the L4-5 level and taped securely in place. The patient is then place in a double envelope of polythene, in a specially constructed bath. Ventilation is maintained through an insulated endotracheal tube, using oxygen enriched air heated to 80°C, which minimises respiratory loss. A urinary catheter and nasogastric tube are inserted and left to maintain continuous drainage during treatment. Three oesophageal and three rectal thermometers are placed at 3 cm spacings, and a tracheal thermometer is also inserted. The electrocardiogram, pulse and temperatures are recorded continuously and intermittent measurements of the blood and central venous pressures taken. The intravenous infusion is heated to 45°C by passage through a heated coil.

The polythene envelope surrounding the patient is heat sealed and low melting point wax (melting point 43-46°C) at 50°C is pumped into the bath, surrounding the enclosed patient. The normal physiological process of heat loss is reversed and energy is introduced into the body at the rate of expproximately 3250 calories per min (Fig. 1), so raising the body core temperature through 5°C over a period of 1 h. When the temperature reaches 41°C all but a thin layer of solid wax is removed. There follows a slower rise in temperature to 41.8°C. Constant temperature to within 0.1°C is then maintained by opening the envelope which encloses the patient and varying the area of skin exposed for the evaporation of sweat, and the thickness of the insulating layer of wax. Narcosis is maintained by intermittent opiates together with a continuous epidural block using 1% lignocaine.

At the end of the hyperthermic period (usually about 4 h) the patient is removed from the bath and cooling is rapid.

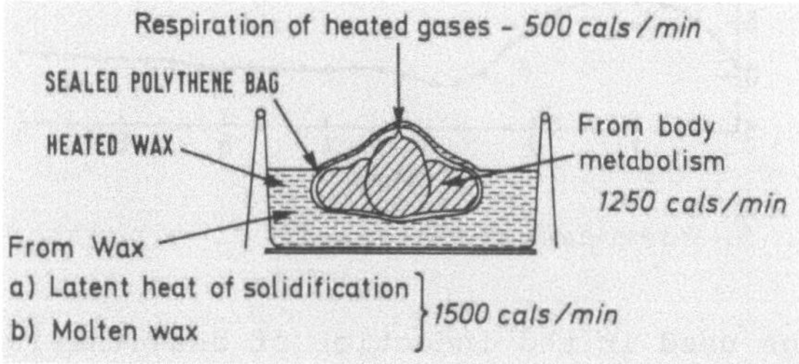

Fig. 1. Sources of heat

4. HAEMODYNAMIC RESPONSE

Patients are closely monitored during treatment. During active heating the temperature in the oesophagus is 1-2°C higher than in the rectum (Fig. 2). Without epidural blockage, the heart rate increases rapidly with temperature (Fig. 3). The increase is linear and averages 8.5 beats/min/°C rise in oesophageal temperature, with a range of 5-15 beats/min/°C. In the same patient, heated on different occasions, the rate of increase in heart rate depends on the rate of heating (Fig. 2). Fast heating initially increases the heart rate less per degree rise in temperature. However, at the end of active heating the heart rate continues to rise for some time, reaching a typical maximum for each patient, independent of the rate of heating, and indicates a limit to the rate at which compensatory mechanisms can respond to stimuli. The final heart rate at stable temperature corresponds to a rise of 11 beats/min/°C^{-1} (range 8-18 beats/min/°C). Stable narcosis is required to maintain this constant heart rate.

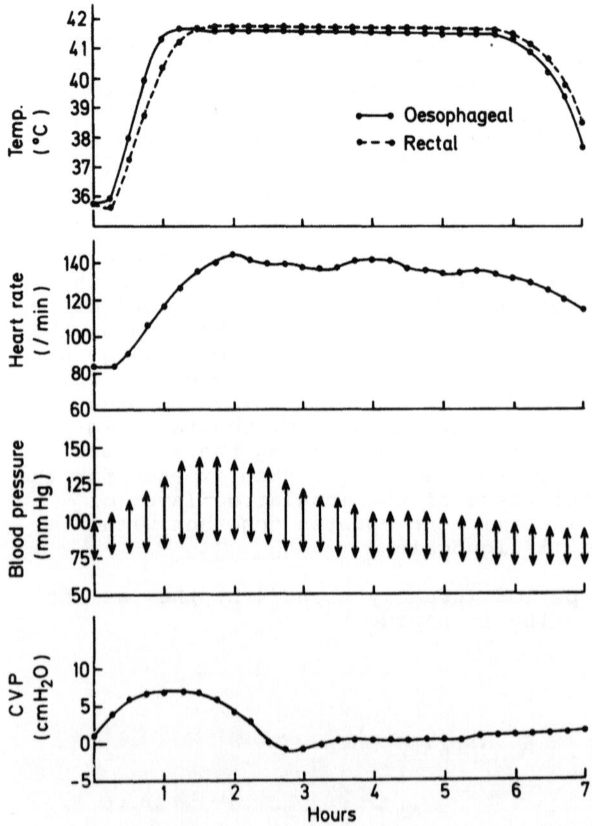

Fig. 2. Haemodynamic response to hyperthermia

Drugs used in the induction of anaesthesia may induce the initial mild degree of hypotension (Fig. 2). Thereafter, during heating, the systolic blood pressure rises by 20-50 mm Hg, with little change in diastolic pressure. At the same time the central venous pressure rises by 5-10 cm H_2O. These changes probably reflect the stress caused by changing temperature. When the temperature is stabilized, both arterial and venous pressures return to the initial values which are then maintained througout treatment.

5. BIOCHEMICAL AND HAEMATOLOGICAL CHANGES

Fluid and electrolyte losses are great during hyperthermia and serum electrolyte concentrations are monitored at half-hourly intervals throughout treatment. Initially fluid and electrolytes were infused according to the serum electrolyte concentration, using the packed cell volume as a measure of dilution. The central venous pressure is an unreliable guide in the slow and continous replacement of fluid required in these vasodilated patients. From these results the intravenous régime recorded in Figure 4 was developed. Mainly sodium chloride and water are given at the beginning of treatment, with potassium being introduced towards the end. Plasma is infused to support

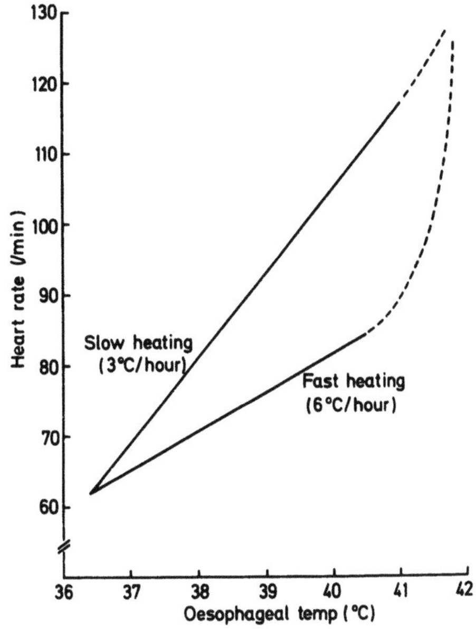

Fig. 3. Variation of heart rate with oesophageal temperature during induction of hyperthermia

blood volume and compensate for the vasodilation which occurs during treatment. This causes no significant change in the serum sodium or chloride concentration during hyperthermia. There is a slight early decrease in the serum sodium concentration, averaging 4mEq/1, which coincides with, and has been attributed to, premedication and anaesthesia (10). The serum sodium and chloride concentrations, measured 24 h after treatment, show no significant change from the pretreatment

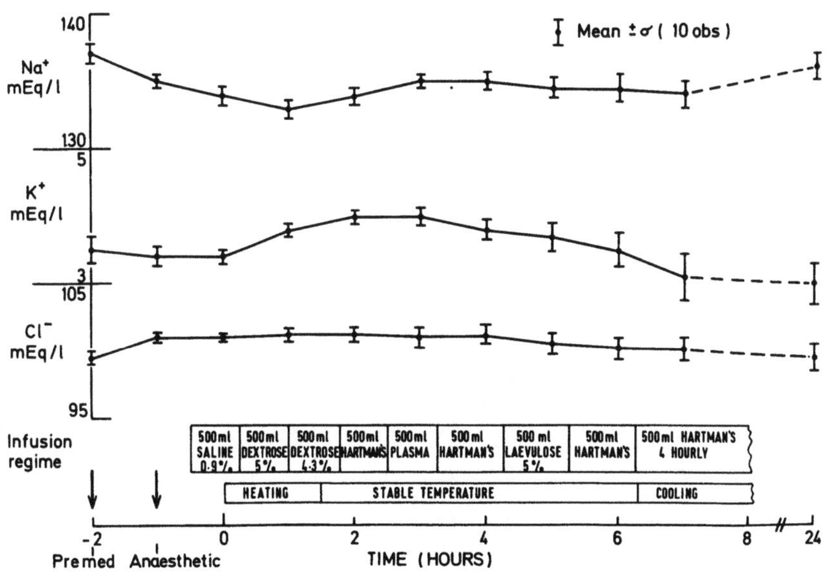

Fig. 4. Plasma electrolytes during hyperthermia

values. The serum potassium does rise by a mean value of 0.5 mEq/1 during the first 90 min of treatment. This increase is not maintained beyond 3 h, when a slow fall in potassium starts, and 24 h after treatment the serum potassium concentration averages 0.5 mEq/1 lower than the pretreatment value.

Losses during treatment were estimated from urinary output and sweat collection data (Table 1). Sweat loss was assessed by enclosing one limb of each patient in a polythene bag and multiplying the volume of sweat collected by the percentage of the total surface area enclosed.

Table 1. Electrolyte and fluid loss and replacement during hyperthermia

	ml/hr	mEq/hr			mEg/1			Na:K
		Na^+	K^+	Cl^-	Na^+	K^+	Cl^-	
Urine[a]	52 o=45%	3.3 o=40%	2.1 o=50%	3.7 o=40%	67 o=30%	51 o=25%	83 o=35%	1:3:1
Sweat[b]	514 o=30%	43 o=40%	2.0 o=20%	41 o=25%	84 o=12%	3.6 o=5%	80 o=12%	23:1
Infusion	570	52	1.0	43	-	-	-	-

[a]mean of 40 measurements.
[b]mean of 60 measurements.

Data given by ALLEN et al. (1) indicates that 60% of sweat is from the arms and legs, which together represent 59% of the body surface area. Half-hourly sweat samples were also collected from the trunk on weighed pads of filter paper to measure sweat electrolyte concentration and variation in sweat rate. Sweating is profuse during hyperthermia, averaging over 500 ml/h. The sweat rate increases rapidely with temperature. At stable temperature, fluctuation of up to 10% occurs in the sweat rate, but there is no further increase during prolonged exposure.

The serum activities of lactate dehydrogenase, aspartate aminotransferase and alanine aminotransferase, all of which have been shown to be stable at the temperatures involved (2), were monitored regularly. In 47 treatments, conducted in 10 patients with non-sensitive tumours, during which neither the rectal nor the oesophageal temperature rose above 41.8°C, there was no significant increase in the serum enzyme activities, and the serum bilirubin concentration rose from an average of 0.47 mg/100 ml to an average of 0.62 mg/100 ml, no value lying outside the reference range (Table 2). However, during 17 treatments within the same group, when the temperature was held between 41.8°C and 42.2°C for periods of 10-40 min, the average serum activity of aspartate aminotransferase increased by a factor of 25, that of analine aminotransferase by a factor of 8 and the serum bilirubin concentration rose to 1.56 mg/100 ml. Raised serum enzyme activity has also been found in animals heated above 42°C (2) and a temperature of 41.8°C was therefore taken as the safe maximum for therapy. In patients with sensitive tumours there are frequently large increases in serum enzyme activity following treatment, which together with a rise in the uric acid level may represent effects of tumour necrosis.

Table 2. Serum enzyme and bilirubin concentrations

	normal range	before heating	after heating not above 41.8°C	after heating above 41.8°C
L.D.H. i.u./l	48-265	181	178	546
A.S.T. i.u./l	4-20	12	13	278
A.L.T. i.u./l	2-17	10	12	78
Bilirubin mg/100 ml	o.2-0.8	0.47	0.62	1.56

During the hyperthermic period the haemoglobin concentration varies bout the mean value by up to 0.5 g/100 ml. The haemoglobin concentration falls by an average of 0.8 g/100 ml within 36 h of each treatment, following which there is a reticulocytosis of 3-4%, leading to partial recovery, but initial values are not attained until the end of the treatment course. This may be due to the thermal destruction of ageing red cells. Similar results were found by KARLE (8) in induced fever in rabbits. Serial bone marrow samples, taken before and after each treatment, have shown no morphological damage, and together with the subsequent reticulocytosis suggest that hyperthermia does not cause marrow depression. During heating there is a leucocytosis, the average leucocyte count rising from 8.2 x 10^3/mm^3 at 37°C to 13.7 x 10^3/mm^3 after 4 h at 41.8°C. It returns to normal within 24 h. This is similar to the changes caused by stress, when there is a discharge of additional white cells into the circulation.

6. CLINICAL RESULTS

67 patients in the terminal stages of cancer, recurrent following conventional therapy, have been treated by whole body hyperthermia. Each patient was considered unsuitable for further treatment by conventional means.

Patients in all age groups, from 3-75 years, have been treated, and in 272 treatment sessions a total of 665 hours over 40°C has been given. Patients tolerate hyperthermia well; the maximum number of treatments in any one patient being 24.

49 patients have been treated with hyperthermia alone; 15 with hyperthermia in combination with cytotoxic drugs and three in combination with local radiotherapy. 21 patients were excluded from the series. 14 were treated in the developmental stages of the method, when temperatures above 40°C were not routinely used. These patients did not respond and it is now accepted that temperatures in excess of 41°C are needed (5). A further 7 patients, with no obviously measureable tumours, were treated for symptomatic relief of pain only.

7. ASSESSMENT OF RESPONSE

The response to treatment was judged favourable if there was weight gain, increased mobility or pain relief together with objective evidence of regression in tumour size on direct measurement, pathological evidence of necrosis in serial biopsy specimens or radiological evidence of regression. Further evidence of heat induced tumour necrosis was obtained at necropsy in five of the six patients who died soon after hyperthermia.

8. HYPERTHERMIA ALONE

Initially all patients were treated over a 4-h period above 41°C, at weekly intervals; later patients were fractionated over three treatments. Though the numbers were small, tumours of gastro-intestinal origin and sarcomas appeared to respond more than genito-urinary or breast neoplasma: lung tumours and malignant melanomas showed an intermediate response (Table 3).

9. CASE REPORTS

Sarcoma. Eight cases. Six were soft tissue sarcomas, of which four responded favourably to treatment. In one patient a lung deposit disappeared; a second showed healing of pathological fractures and a third showed complete regression of a fibrosarcoma recurrent in the operation scar. The fourth patient, with an advanced liposarcoma, died 24 h after treatment. Necropsy showed recent massive necrosis throughout the tumour (Fig. 5). A child with a rhabdomyosarcoma showed no response and one fibrosarcoma had subjective improvement only. Two patients with osteogenic sarcoma were treated; the first had a hindquarter amputation - 2 months after the end of an 18 month course of treatments the tumour recurred in a heat resistant form. The second showed no response to treatment.

Carcinoma of the Stomach. Four cases. Two anorexic patients who had been in great pain gained weight and were able to lead a relatively normal life. The third had extensive mediastinal and lung metastases and died 48 h after treatment. At necropsy there was extensive necrosis of the tumour causing compression of the bronchus. The fourth had pain relief and weight gain, but died suddenly 1 month after treatment from a pulmonary embolus. Necropsy was not performed.

Carcinoma of the Colon. Seven cases. One patient had almost complete regression of massive hepatomegaly; a second had regression of skin nodules in the lower abdomen. One further patient had regression in tumour size; the fourth died 3 weeks following hyperthermia. At necropsy there was very extensive tumour necrosis. The other three cases showed no response.

Malignant Melanoma. Eight cases. In four there was good initial regression of secondary deposits and pain relief. One patient had pain relief alone and three showed no response.

Carcinoma of the lung. Four cases of bronchogenic carcinoma. In one patient with adenocarcinoma there was regression of the primary lung tumour (confirmed at necropsy) (Fig. 6), though the secondary deposits remained active. The second patient, with a squamous carcinoma, showed regression of a secondary deposit in the lumbar spine; the third had relief from pain but no tumour regression. The fourth died within 24 h

Table 3. Results of treatment with hyperthermia alone

Tumour type	No. of patients treated	Previous treatment Surgery	Radiotherapy	Chemotherapy	Objective responses	Subjective response	No response	Survival from start of thermotherapy (weeks)
Sarcoma	8	6	4	2	3 + 1[a]	6	1	84, 52, 20, 8, 4, 2, 1, 0.14
Gastric carcinoma	5	5	0	1	3 + 1[a]	3	1	16, 16, 0.28, 12, 4+
Carcinoma colon	8	8	0	1	4	3	3	20, 8, 4, 0.71, 10, 13, 4, 4
Melanoma	8	8	2	2	3	5	3	12, 12, 8, 8, 4, 4, 3, 2
Carcinoma lung	4	0	3	0	2	3	1[a]	24, 20, 4, 0
Carcinoma breast	2	2	2	1	0	1	1	12, 12
Ovarian carcinoma, teratoma testes	4	4	2	3	0	0	4	24, 16, 8, 4
Neuroblastoma, Nephroblastoma	3	3	3	3	1 + 2[a]	1	0	24, 0.85, 0.28
Miscellaneous	7	4	5	6	2	2	4	32, 24, 12, 0.57, 10, 24+, 6

[a]Necropsy evidence of recent tumour recrosis.

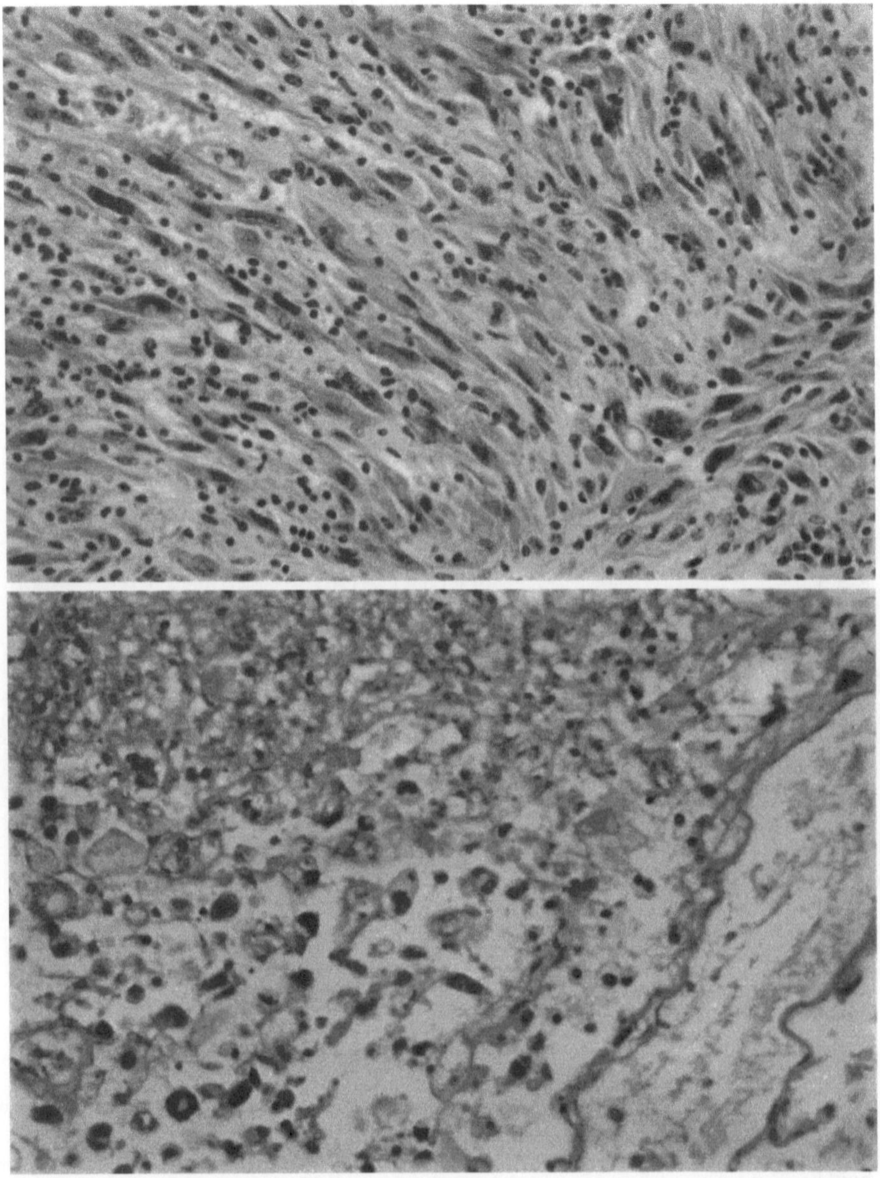

a)

b)

Fig. 5. Sarcoma. a) Before treatment. b) After 4-h hyperthermia. H & E, x 270

of hyperthermia; prior to death there was evidence of disseminated intra-vascular coagulation and at necropsy there was extensive haemorrhage into the tumours.

Ovarian and Testicular Tumours. Two testicular teratomas and two ovarian papillary tumours showed no response to treatment.

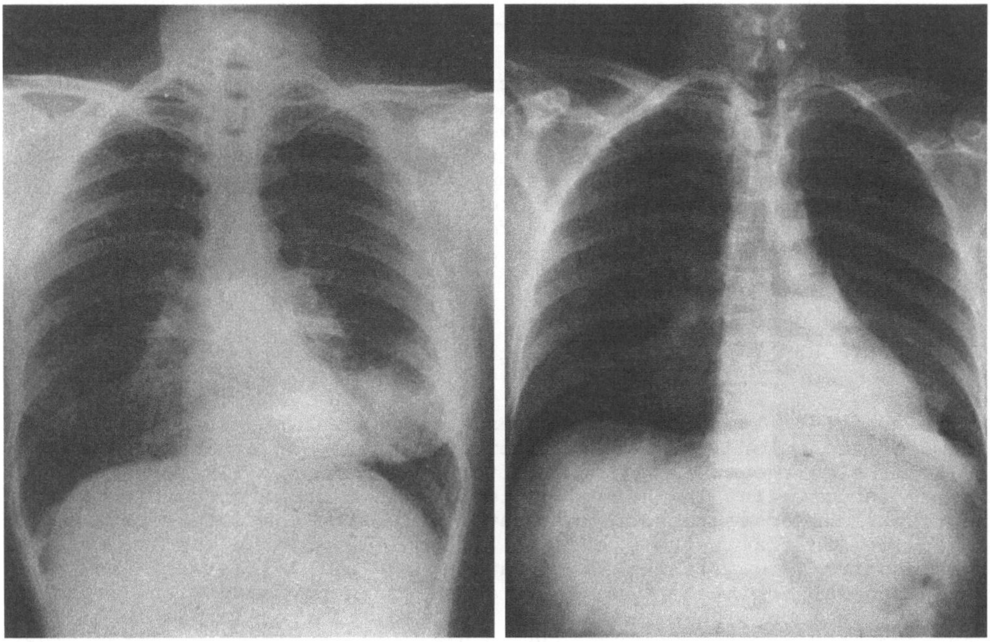

a) b)

Fig. 6. Adenocarcinoma. a) Before treatment. b) After 10-h hyperthermia

Neuroblastoma and Nephrobastoma. Two children with neuroblastomas were treated. One showed a good initial response, with healing of ulcerated skin over the tumour in his jaw. The second showed initial improvement till he developed respiratory difficulties and died 2 days after treatment. Necropsy showed multiple haemorrhagic areas of necrosis in the tumour. One child with a nephroblastoma showed initial improvement but also died from a respiratory arrest. At necropsy there was gross necrosis of the tumour.

Breast cancer. Two patients with scirrhous carcinomas were treated, one obtaining pain relief alone.

Miscellaneous Tumours. One case of mycosis fungoides showed initial healing and there was pain relief alone in a case of adenocarcinoma of the nasopharynx. One case each of chronic myeloid leukemia and transitional cell carcinoma of the bladder showed no response.

10. HYPERTHERMIA IN COMBINATION WITH CYTOTOXIC DRUGS

Hyperthermia was given in three fractions, at weekly intervals. The first lasting 90 min, above 41°C, the second two fractions being of 4 h. Cytotoxic drugs were given by intervenous bolus injection, during the last treatment. Patients with malignant melanoma were given melphalan 1 mg/kg; the others were given cyclophosphamide 200 mg during the period of temperature rise and fluorouracil 15 mg/kg with vincristine 1 mg at a temperature of 41°C. The results in the 15 patients given cytotoxic drugs during hyperthermia are shown in Table 4.

11. CASE REPORTS

Gastrointestinal Tumours. Eight cases. One patient had an adenocarcinoma of the colon with large hepatic metastases enlarging the liver to 12 cm below the costal margin. After treatment the liver regressed to 5 cm subcostally. The second patient, with a cholangiocarcinoma of the liver, showed regression of hepatomegaly with complete clearance of jaudice (initial serum bilirubin 12.2 mg/100 ml) and resolution of gross ascites (Fig. 7). The third patient had an undifferentiated carcinoma and was admitted to hospital as an emergency case with large bowel obstruction. There was a mass 20 cm in diameter in his left iliac fossa and he had renal failure due to ureteric involvement. After treatment he had complete regression of the mass with a return of normal renal and bowel function. A fourth patient showed regression in lower abdominal secondary deposits and the fifth some regression of hepatomegaly. The sixth patient showed necrosis on serial biopsy of the tumour. There was symptomatic improvement in one patient with adenocarcinoma of the gall bladder and no response in a second patient with a similar tumour.

Malignant Melanoma. Three metastatic cases. Two patients showed regression of involved axillary nodes and the third showed regression of hepatomegaly, though her secondary nodules did not change in size.

Breast Cancer. Three cases. These patients had cancer en cuirasse and were in severe pain. In the first two there was relief of pain enabling opiates to be discontinued, and regrowth of skin over the tumours. One of these patients died within 12 h of a further treatment, given for recurrence 3 months later. There appeared to be no pathological evidence of tumour necrosis and death was attributed to disseminated intravascular coagulation. The third patient died 48 h after treatment with disseminated intravascular coagulation. In this case necropsy showed evidence of recent cell death in the tumour metastases, which involved liver, adrenal, both kidneys, skull, uterus, pancreas, vertebral and dura.

Miscellaneous. One case of osteoblastoma was treated. Though the patient had pain relief there was no regression of the tumour.

Table 4. Irradiation combined with whole body hyperthermia: Three osteogenic sarcoma

Tumour type	No. of patients treated	Previous treatment			Objective response	Subjective response	No response	Survival from start of thermotherapy (weeks)
		Surgery	Radiotherapy	Chemotherapy				
Gastrointestinal tumours	8	6	1	3	6	6	0	32, 20, 20, 16, 12, 8, 16, 10
Carcinoma breast	3	3	3	2	2 + 1[a]	2	0	20, 12, 0.28
Melanoma	3	3	0	0	3	1	0	52, 44, 24
Osteoblastoma	1	1	0	0	0	1	0	12

[a]Necropsy evidence of recent tumour necrosis.

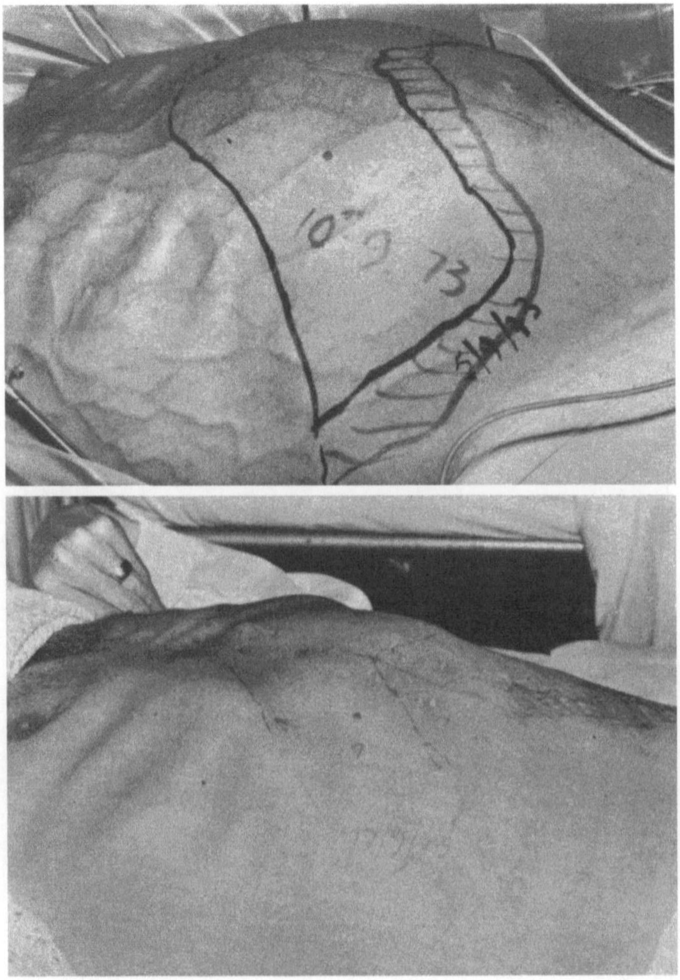

Fig. 7. Resolution of gross ascites in cholangiocarcinoma of the liver. a) Before treatment. b) 1 week later, after two hyperthermic sessions

12. HYPERTHERMIA COMBINED WITH LOCAL RADIOTHERAPY

Three patients with osteogenic sarcomas were treated by local radiotherapy immediately following 4 h at 41.8°C (Table 5). In two cases of disseminated disease individual lung metastases were irradiated with single dose of radiotherapy. In these cases the irradiated lesion regressed to a larger extent than would have been expected by radiotherapy alone. In both there was an arrest of other non-irradiated metastases, but no regression in size. One patient with a fulminating osteogenic sarcoma of the leg, with no metastases, was treated with 5500 rads in daily fractions over 20 treatments, and weekly hyperthermia followed immediately by a radiotherapy fraction. The tumour regressed, with re-ossification of bone and complete loss of pain. (Fig. 8) The skin remained intact and there was no evidence of metastases at 6 months.

Table 5. Results of treatment with hyperthermia in combination with chemotherapy

	1	2	3
Treatment of primary	Surgical amputation	Irradiation	Irradiation while at peak temp. (hours at 41.8)
Present state of primary site		Fixed immobile	Full range of movement now present
Secondaries	Bilat. pulmonary	Bilat. pulmonary	NIL
Treatment of secondaries	Thermotherapy + irradiation to one large pulmonary secondary	Thermotherapy + irradiation to one large pulmonary secondary	
Present findings	The one secondary reduced in size others progressing rapidly	The one secondary reduced others unchanged. Patient continues well	Patient well

13. COMPLICATIONS

Complications may arise from the method, from the physiological response to high temperatures or from the toxic effects of tumour breakdown and absorption. Considering that patients have been maintained in an unconscious state at temperatures over 40°C for 665 h, complications of a major nature have been markably few. In 1966, before the present method was in use a patient developed ventricular fibrillation and died. This was due to her temperature reaching 43°C as a result of thermometer failure and was the only fatality directly attributable to induced hyperthermia. The importance of several thermometers has been stressed; these should be placed at different levels in the rectum and oesophagus and read out on at least two separate systems. The highest reading thermometer is taken as the true reading.

Half of the patients developed circumoral herpes simplex following the first treatment, but not after subsequent treatments. As there is no evidence of immune supression during treatment it may be that the virus is extruded from heated cells and becomes active or recognised by the host.

Sore throats, pressure sores due to prolonged immobilisating during treatments, and superficial burns in oedematous, hypoproteinaemic patients may be minimised by adequate padding and an air mattress. For grossly oedematous patients the wax temperature is decreased to 45°C.

Five adult patients died within 48 h of hyperthermia. Their deaths were associated with evidence of disseminated intravascular coagulation. In three necropsy showed recent tumour necrosis. This is a reported complication of agressive chemotherapy in the presence of widespread sensitive tumour and indicates the necessity of fractionating hyperthermia in the presence of disseminated sensitive tumour.

Two children with advanced neoplastic disease died shortly after treatment. They were given opiates for the relief of distress and died of respiratory complications. Another patient died of fibrosing alveolitis, possibly due to repeated exposure to the hot, moist, ventilating gases

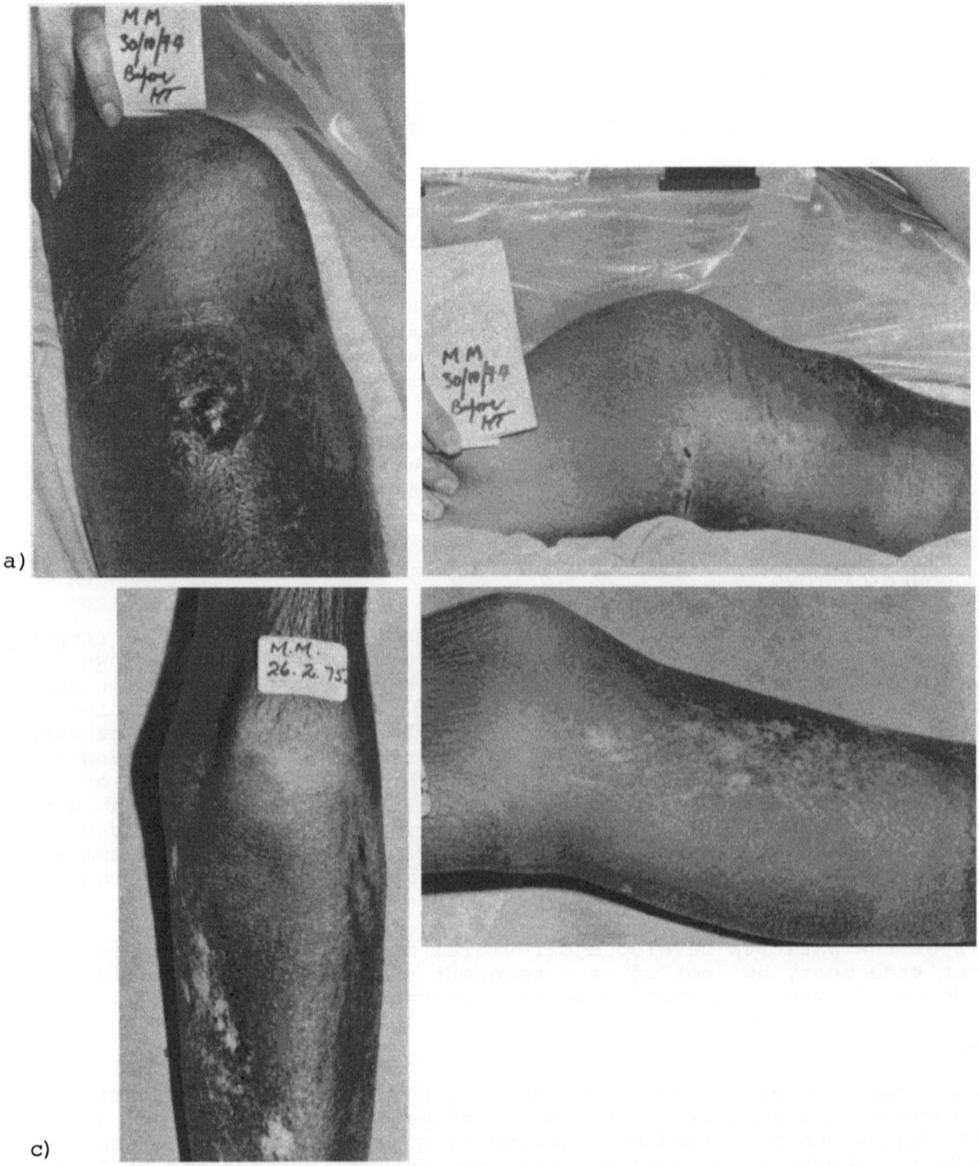

Fig. 8. Osteogenic sarcoma. a, b) Prior to treatment. c, d) 2 months after treatment by irradiation with whole body hyperthermia at 42°C

then in use (6), or as a result of treatment with bleomycin 6 months previously. Hot dry gas is now used and this complication has not recurred.

FRACTIONATION OF HYPERTHERMIA

The effect of hyperthermia is to stress the cellular enzymes necessary for life. Some tumour cells respond less well to this stress than the normal cell and die. Those tumour cells which survive are more resistant to further heat therapy. In disseminated disease there is a large tumour mass which if treated vigorously, necroses en masse, liberating toxic products into the circulation, and may cause disseminated intravascular coagulation. Treatment must therefore be fractionated into doses which cause a similar amount of tumour necrosis. In practice fractions are given at 2 h, 4 h and 6 h on successive weeks. Patients with a sensitive tumour show evidence of a systemic reaction following the first treatment with a decrease in platelet count. A second fraction of the same period causes no upset, but prolongation of the fractions again causes a reaction. This reaction usually last for up to 48 h. Patients with unresponsive tumours have no such reaction.

The addition of cytotoxic drugs during the hyperthermia period potentiates the effect of hyperthermia. If alkylating agents are given at the beginning of active heating the normal dose must be decreased as marrow depression occurs. At peak temperature the normal dose of drug is possible without causing severe marrow depression.

Radiotherapy also potentiates hyperthermia, by causing breaks in the cell which are not repaired because of the disruption of repair enzymes caused by hyperthermia. In whole body treatment, local radiotherapy may be an effective method for treating the main tumour mass.

15. FUTURE MANAGEMENT

By an extrapolation of animal studies a single treatment in excess of 20 h at 42°C would be needed to produce total tumour necrosis (7). This treatment period may become possible with a greater understanding of the physiological processes which occur at 41.8°C. For each 1°C rise in temperature above 40°C the time required for the same percentage of tumour necrosis is halved. Local heating of the tumour mass to a higher temperature, during whole body treatment at 41.8°C, may be possible and shorten the time required. There is need for considerable work to be done to assess the exact thermal sensitivity of various tumours and this should be recorded in units related to the time required over 40°C to cause 50% tumour necrosis. Potentiating agents could then be described in terms of this factor and the treatment fractions calculated exactly.

16. SUMMARY

At the present time, whole body hyperthermia has been used to palliate disseminated disease in terminally ill patients, and to increase the quality of their lives by relieving tumour induced pain. Soft tissue sarcomas and tumours of gastro-intestinal origin may prove to be sensitive to hyperthermia. The future depends on finding individual tumour sensitivities and fractionating treatment more effectively.

ACKNOWLEDGEMENTS

We wish to thank the British Medical Journal and the Lancet for permission to publish the illustrations shown in this article. Also, The Melville Trust for cancer research for funding the project. We would like to thank all referring consultants for their co-operation and help in the treatment of these patients.

REFERENCES

1. ALLAN, J.A., ARMSTRONG, J.E., RODDIE, I.C.: The regional distribution of emotional sweating in man. In: Proc. physiol. Soc. 1973, p. 37
2. BURGER, F.J.: The effect of hyperthermia on serum enzymes activity. S. Afr. med. J. 44, 889-901 (1970)
3. DICKSON, J.A., ELLIS, H.A.: Stimulation of tumour cell differentiation by elevated temperature (42°C) in rats with transplated Yoshida tumours. Nature (Lond.) 248, 354-358 (1974)
4. FINDLAY, J., McMANUS, J., LUDGATE, C.M., PETTIGREW, R.T.: The effects of hyperthermia therapy on the liver. II. Morphological observations. J. clin. Path. (In press, 1975).
5. GIOVANELLA, B.C. LOHMAN, W.A., HEIDELBERGER, C.: Effect os elevated temperatures and drugs on the viability of L. 210 Leukaemia cells. Cancer Res. 30, 1623-1631 (1970)
6. HENDERSON, M.A., PETTIGREW, R.T.: Induction of controlled hyperthermia in the treatment of cancer. Lancet (1971), 1275-1277
7. JOHNSTON, H.J.: The action of short radiowaves on tissues - A comparison of the thermal sensitivities of transplantable tumours in vivo and in vitro. Amer. J. Cancer 38, 533-550 (1970)
8. KARLE, H.: Destruction of red blood cells during experimental fever. Acta med. scand. 186, 349-359 (1969)
9. McKENZIE, A.: Personal communication, 1975
10. STEVENSON, D.E.: Chanoes caused by anaesthesia in the blood electrolytes of the dog. Brit. J. Anaesth. 32, 353-363 (1960)

Chapter 7
Hyperthermic Perfusion of Extremities for Melanoma and soft Tissue Sarcomas

J. S. STEHLIN, JR., B. C. GIOVANELLA, P. D. DE IPOLYI, L. R. MUENZ, R. F. ANDERSON, and A. A. GUTIERREZ

1. INTRODUCTION

Since the work of KLOPP (8, 9), CREECH (3-6) and our group (13-16), perfusion has attracted wide attention as a method of treating patients with certain types of malignant neoplasms of the extermities. We began normothermic perfusion in 1957.

A. Normothermic Perfusion

The lower extremity may be perfused through three levels: popliteal, femoral, and iliac (12). For popliteal perfusion, the knee is flexed 45° and a short longitudinal incision is made along the medial aspect of the lower tigh. A tourniquet is applied around the midthigh.

The most suitable level for normothermic perfusion of the entire lower extremity is through the external iliac artery and vein. These vessels are exposed extraperitoneally through a short oblique incision which divides the rectus muscle, and the tips of the catheters are passed to a point just distal to the inguinal ligament. An Esmarch bandage is employed as a tourniquet, being wrapped tightly around the upper portion of the thigh and held in this position by a steinmann pin driven into the anterior superior spine of the ilium. It is important that the major collateral vessels, i.e., deep circumflex iliac, inferior epigastric, and obturator vessels, be occluded or ligated to reduce leakage.

For melanomas of the lower extremity which have metastasized to the femoral lymph nodes, perfusion at the femoral level is performed as a second-stage operation subsequent to iliac perfusion. In such cases, femoral perfusion is immediately preceded by femoral lymphadenectomy. Because of its simplicity, femoral perfusion may also be employed as a primary procedure for melanomas of the lower extremity. A short vertical incision is placed high in the femoral triangle. The saphenous vein is ligated and its stump is used to catherize the superficial femoral vein. The Esmarch bandage is applied as for iliac perfusion.

If radical axillary dissection is not anticipated, the upper extremity is perfused via the axillary artery and vein. These vessels are exposed through a short infraclavicular incision. The tips of the catheters are passed to the region of the upper third of the humerus and a proximal tourniquet is applied.

If axillary dissection is indicated, the incision is begun at the medial end of the clavicle and curved downward and outward to the posterior axillary line. Axillary dissection followed immediately by perfusion is carried out. At present, no satisfactory tourniquet is available for perfusion of the axillary contents as well as the entire upper extremity.

In arriving at the ideal dose of the chemotherapeutic agent, the two most important limiting factors are (1) local tissue tolerance, and (2) the leakage factor. The local tissue tolerance is indicated by the amount of drug which the normal tissues can tolerate without being permanently damaged. The leakage factor refers in general to the extent of cross-circulation, expressed in percentages, between the perfused region and the remaining portion of the body (14).

A technique for the continuous monitoring of leakage was developed by us in 1959 (13). The method involves the use of a scintillation probe, rate meter, and a rectilinear recorder. Radioactive iodinated serum albumin has been employed in preference to chromium-tagged red cells, because of the relative ease with which it may be handled and the almost uniformly satisfactory results obtained. We now are able to determine how efficiently the perfusion circuit has been isolated from the remainder of the body before the chemotherapeutic agent is injected. Thus, the proper dosage of the compound may be determined and complications incident to leakage reduce to a minimum.

Virtually no leakage is associated with perfusion of the distal extremities; the only consideration in determining the proper dose of the chemotherapeutic agent to be delivered to these regions is local tissue tolerance. In the iliac perfusion to leakage usually varies from 10% to 30% at the end of 1 h (12).

During the past 15 years, regional perfusion, for some surgeons, has become acceptable therapy for certain cancers of the extremities. In our program, the same chemotherapeutic agent, that is, melphalan, has been employed consistently. Since little was known about the procedure in the beginning, use of perfusion was limited to locally advanced tumors, for which amputation was necessary, or to disseminated cancer. As more general information was obtained and good objective responses observed after perfusion, its clinical use was extended to include tumors considered possibly to be curable.

B. Hyperthermic Perfusion

For many years, investigators have commented on the destructive effect of heat upon cancer cells. CAVALIERE et al. (1) reported biochemical and clinical studies which demonstrated selective sensitivity of cancer cells to heat. Twenty-two patients with cancers of the extremities in that study were treated by limb perfusion with heated blood without chemotherapy. Marked regression of certain tumors, especially melanomas, was observed. Based upon their work, a major change in our technique was instituted in 1967, namely, the addition of hyperthermia. At first our methods of heating the blood and warming the extremity were rather inadequate. Blood was heated by an inexpensive 250-Watt infrared lamp and the skin warmed by moist towels. Now the blood is heated by a closed circuit water-circulating unit, connected to a heat exchanger, with a temperature range of 50-150° F (10-65°C). Additional heat is applied externally by wrapping the extremity with Aquamatic K-Pads. The temperature of the perfused blood, the skin, muscle, and, when possible, tumor, are continuously monitored.

A striking improvement in the objective response of melanomas was noted when heat was added to the perfusion system without changing the chemotherapeutic agent (11).

2. MELANOMAS

A. Technique

Our perfusion technique has remained essentially the same since 1957, except for the addition of heat. Melphalan has been used consistently, although the amount has been lowered with the advent of hyperthermia. For the lower extremity, the temperatures of the skin and muscles of the calf and lower medial portion of the thigh are monitored continuously by thermistor probes. For the upper extremity, probes are applied to the skin and muscles of the dorsal surface of the forearm. When gross tumor is present, the temperature is monitored by a probe inserted directly into the tumor. Additional heat is applied externally by wrapping the extremity with an Aquamatic K-Pad. Perfusion is begun with the temperature of the blood at 110.0°F (43.3°C). When the temperature of the skin and muscles reaches 102.0-104.0°F (38.8-40.0°C), melphalan is injected into the venous line. At this point, the temperature of the blood is adjusted to maintain the desired level in the skin and muscles. It is monitored every 2 min and recorded every 5 min throughout the procedure. The temperature and duration of perfusion vary according to the extent of melanoma within the limb and the condition of the peripheral vessels. At the conclusion, the extermity is washed out with a solution of low molecular weight dextran. The risk of serious injury to normal tissues is directly proportional to the amount of drug used, the temperature obtained, the duration of perfusion, and the condition of the peripheral vessels. We prefer not to leave the tourniquet intact for more than 2 1/2 h. The vast majority of our patients are perfused only once, although perfusion is repeated if there is extensive, persistent, or recurrent disease. One patient has undergone four such procedures.

B. Major Complications

During the past 7 years, 185 hyperthermic perfusions have been performed on 165 patients. Two patients (1.1%) died as a result of the operation. One was a 75-year-old female who died of renal failure and the other a 67-year-old female who died from a pulmonary embolus; both had extensive in-transit disease. Tissue necrosis requiring major amputation was encountered in three patients (1.6%), and was severe enough to require débridement without amputation in two additional patients. One patient suffered fottdrop after prolonged popliteal perfusion with the tourniquet in the middle part of the thigh. Although seen in the early phases of our work, renal complications have not been observed during the last 5 years.

C. Classification of Melanoma

We have used the following classification of melanoma for many years. Stage I tumors have no clinical signs of metastasis or any microscopic evidence of melanoma in the nodes removed at regional lymphadenectomy. In stage II, local recurrence occurs within or adjacent to the scar of a primary lesion. In stage IIIA, regional intransit metastasis or satellitosis develops without clinical or microscopic evidence of metastasis in the regional lymph nodes. Stage IIIB consists of proven metastasis to the regional lymph nodes with no evidence of in-transit disease. In stage IIIAB, in-transit metastis occurs in addition to metastasis to the regional lymph nodes. Stage IV tumors are those with distant metastasis. This group also includes primary tumors of the upper extremity, excluding the shoulder region, with metastasis to the supraclavicular lymph nodes, and primary tumors of the lower extremity with metastasis to the iliac and/or obturator lymph nodes. The number of patients in each stage is shown in Table 1.

Table 1. Hyperthermic perfusion for melanoma of the extremities

Stage	No. of patients
I	70
II	8
IIIA	30
IIIB	17
IIIAB	18
IV	22
Total	165

From STEHLIN et al. (17).

D. Outline of Treatment

Perfusion is omitted for patients whose primary lesion arises in the deltoid region or on the proximal part of the thigh. For patients in clinical stage I whose primary tumor arises distal to the insertion of the deltoid muscle, an axillary dissection is carried out, preserving the pectoral muscles. This is followed immediately by perfusion. For those primary melanomas that arise distally from the middle portion of the thigh, perfusion is performed through either the femoral or iliac vessels; elective lymphadenectomy is omitted.

The primary lesion is identified histopathologically according to the classification of CLARK and FROM (2). For lesions at level I and level II, simple excision is performed without perfusion. The vast majority of our patients are referred to us after excision of the primary lesion. As a rule, we excise the scar, usually with closure of the wound. Although we do not hesitate to apply a skin graft at the time of perfusion, this has been necessary in only a small percentage of these patients. For patients who are referred with an intact primary melanoma which clinically appears deeply invasive, a small incisional biopsy is taken to confirm the diagnosis and to determine the depth of invasion. Because of the poor results obtained with conventional therapy for lesions which involve the reticular dermis or subcutaneous fat, the limb is perfused, and excision of the tumor is delayed until maximum regression has occured. During the past 4 years, we have injected bacillus Clamette-Guerin directly into the melanoma immediately after perfusion. This same type of adjuvant therapy has been administered to some patients with intransit metastasis. For persistent or recurrent in-transit disease, superficial radiation therapy to small fields occasionally supplements perfusion.

E. Survival Rates and the Biologic Behavior of Stage I Melanoma

The 5-year projected survival rates are calculated according to the Berkson-Gage method.

Stage I. For 70 patients in stage I, the 5-year survival rate is 83.5%. Of these 70 patients, 15 had lesions arising on the upper extremity; 14 of the 15 underwent elective axillary dissection, with all nodes proved negative microscopically. The remaining 55 patients in stage I had lesions arising on the lower extremity; only 2 of these had prophylactic dissection of the lymph nodes. The femoral lymph nodes in the remaining 53 patients were judged clinically negative.

It is interesting to follow the biologic behavior of melanoma in the patients in whom regional metastasis developed after treatment for stage I disease. Ten of the 70 patients died, 4 of them of causes unrelated to melanoma. Six patients, 3 of whom never demonstrated clinical involvement of the regional lymph nodes, died of melanoma. One patient had -.pid enlargement of the femoral and iliac nodes as the initial manifestation of explosive disease and died 10 weeks after node dissection. In the remaining 2 patients, the femoral nodes became clinically positive 19 and 28 months after perfusion; these patients died 9 and 21 months, respectively, after lymphadenectomy. In 4 patients, clinically positive femoral lymph nodes developed after perfusion; they underwent therapeutic dissection and are apparently free of disease at 36, 54, 72, and 85 months. Following perfusion, in-transit metastasis developed in 2 patients and advanced from stage I to IIIA. Both patients underwent a second perfusion; 1 is now living with distant metastasis without regional recurrence, and the other is apparently free of disease 5 years after the initial perfusion. In no patient has the melanoma advanced from stage I to stage II, local recurrence.

Stage IIIA. Two of the 30 patients in this group had lesions involving the upper extremity; their lymph nodes were microscopically negative after axillary dissection. Twenty-eight patients had in-transit metastasis involving the lower extremity. Sixteen of the 28 (57%) had previously undergone lymphadenectomy elsewhere, with the nodes proved microscopically negative. The femoral lymph nodes in the remaining 12 patients were clinically negative, but not proved so microscopically. The 5-year survival rate from the time of perfusion is 71.8%.

Stages II, IIIB, and IIIAB. When stages II, IIIB, and IIIAB are viewed separately, the number of patients in each category is too small for significant evaluation of survival time. However, when all 73 patients with regional recurrences or metastases clinically confined to an extremity are combined, the 5-year survival rate is 48.2%.

3. DISCUSSION

Shortly after heat was added to the perfusion with melphalan, it became obvious to us that the objective response of the melanomas was immensely improved, from 35% to 80%. Whether or not this dramatic increase in objective response would be reflected in an increase in survival rate could not be ascertained until enough time had elapsed. Thus, we elected to wait 7 years from the first hyperthermic perfusion before evaluating the data. Insofar as patients in stage I are concerned, a 5-year survival rate of 83.5% seems good. This is especially true when one considers that 55 of the 70 patients in stage I had lesions arising on the lower extremity, and of the 55, only 2 had prophylactic node dissection. Had all 55 patients undergone lymphadectomy, the nodes in some would have been found to contain melanoma, thus advancing these patients from stage I to stage IIIB.

The 5-year survival rate of 48.2% for the combined 73 patients with recurrences or metastases clinically confined to an extremity appears to be surprisingly high. Presently, we have no accurate way to compare these survival rates with thise of other authors because of lack of conformity in classifications. McNEER and CANTIN (10), in an extensive review of local failures in the treatment of melanoma, said: "Once local failure develops after definitive treatment for melanoma, the prognosis, while not hopeless, is grave." Their reported 5-year survival rate following definitive treatment for recurrences clinically confined to the extremity in 64 patients was 20%.

Since our first hyperthermic perfusion was performed in 1967, we have believed that the merits of the procedure would be reflected most accurately in the results obtained by treatment of patients in stage IIIA, in-transit metastasis without clinical or microscopic involvement of the regional lymph nodes. The 5-year survival rate following hyperthermic perfusion for 30 patients in stage IIIA is 71.8%, which, by any reasonable criteria, is extraordinarily good.

These exceptional results prompted us to conduct an extensive retrospective study of survival times of patients in stage IIIA who were treated prior to the introduction of hyperthermic perfusion. Our report (17) on in-transit metastastic melanoma of the extremities included results obtained following treatment of patients in stage IIIA. In this paper, we wrote: "Regardless of the condition of the nodes, however, in-transit metastases carry a grave prognosis; only a few patients survive five years following any type of treatment." In this group of patients, the projected 5-year survival rate after the initial appearance of in-transit disease was 18.7%. A comprehensive review of these patients has been conducted, and the pertinent information now has been brough up to date. The crude 5-year survival rate of this group of patients reported earlier actually is 22.2%. Prior to 1966, 27 patients in stage IIIA were treated; since 1967, 30 patients in the identical stage have been treated with hyperthermic perfusion. To achieve a more meaningful comparison, the survival rates for both groups of patients have been calculated from the time of initial definitive therapy for stage IIIA disease (Fig. 1). Comparative data are shown in Table 2.

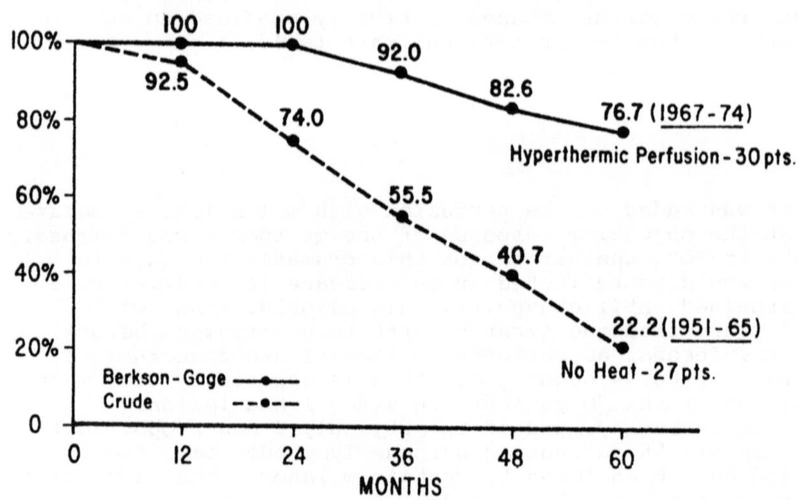

Fig. 1. Survival rates from the time of initial treatment of patients with stage IIIA disease, in-transit metastis. (From STEHLIN et al. (17))

Table 2. Comparative data on stage IIIA metastatic melanoma of the extremities

	No heat (1951-65)	Heat (1967-74)
Number of patients	27	30
Female	13	24
Male	14	6
Extremity involved		
Upper	7	2
Lower	20	28
Regional nodes proved negative microscopically	22/27-81%	17/30-56%
Patients with more than one recurrent (metastatic) nodule	21-78%	23-77%
Radical amputation	8/27-30%	0
Patients perfused	21/27-78%	30/30-100%
Perfusion drug	Melphalan	Melphalan
Average dose lower extremity	1.2-1.6 mg/kg	0.9 mg/kg
Perfusion time	45-120 min	45-120 min
Skin temperature	30-32.2°C	38.8-40.6°C
Muscle temperature	33.3-35°C	38.8-40°C
Five-year survival rate		
Males and females	22.2%[a]	76.7%[b]
Females only	13,30.7%[a]	24,88.7%[b]

[a]Crude.
[b]Berkson-Gage.
From STEHLIN et al. (17).

In the earlier group of 27 patients, treatment consisted of multiple local excisions with or without skin grafting, unheated perfusion with melphalan followed in 7 instances by intra-arterial infusion with various other chemotherapeutic agents, and radical amputation. Amputations were necessary for 6 patients when unheated perfusion either with or without infusion failed to control the disease and for 2 patients when multiple excisions failed to do so. Since 1967, none of the 30 patients who have received hyperthermic perfusion have required amputation.

A comprehensive statistical analysis of the results of perfusion with and without heat was conducted at the National Cancer Institute, based on the length of survival from the time the patient received the first definitive treatment for stage IIIA disease. Forty-eight patients received either unheated or heated perfusion with melphalan, but not both. Nine of the original 57 patients received neither or both of the treatments and, therefore, were not included in this comparative study.

As of June 1974, the median survival time is the same in the two groups: 48 months. However, 22 of the 25 patients still alive are in the group that received the heated perfusate. Hence, the median survival time for the group receiving heated perfuste will soon exceed that of the group receiving unheated perfusate. A crude analysis of the comparative survival experience indicates dramatic superiority for the heated perfusion. The odds of dying in a given month appear to be 5 times greater for the patients who received the unheated perfusion than for those who received the heated perfusion.

Another more detailed analysis was performed, using the Mantel-Haenszel life table approach, taking into account the age and sex of the patient and the length of time from diagnosis to administration of perfusion. The disparity in the numbers of males and females in the two groups had a significant influence on the outcome of this analysis. In the heated group, 21% were males; in the unheated group, 50% were males. Females demonstrated superior survival times in both groups and thus favored the heated group, in which they were more numerous. The waiting time from diagnosis to treatment was of lesser significance in that it was more equitably divided between the two groups; the same was true of the factor of age. Even taking all these factors into account, the odds in favor of the patient who received the heated perfusion are 3 to 1. The probability of encountering such a result purely by chance is only 1 to 10. Another manifestation of the superiority of heated perfusion is seen in the 5-year survival rates. The probability of survival to 5 years is 0.795 for the heated group and 0.344 for the unheated group.

The improved long-term survival of patients with local dissemination of disease within the affected limb is difficult to explain unless some systemic effect of the hyperthermic perfusion is postulated. The elimination of tumor in the extremity cannot be more complete than that obtained with radical amputation. However, long-term survival of patients after amputation for melanoma at this stage is poor; the cause of death is distant metastasis, occult and unrecognized at the time of amputation. If distant metastasis is already present when treatment is initiated for patients in stage IIIA, any improvement in survival time after hyperthermic perfusion must be caused by destruction or inactivation of the distant disease. A plausible mechanism for such effect could be the stimulation of antitumor immunity of the patient by the antigenic material liberated when melanoma is destroyed in loco by perfusion. That such stimulation following hyperthermic perfusion is, indeed, present in at least some patients has been observed clinically. One patient who underwent hyperthermic perfusion for extensive in-transit disease of the extremity also had numerous cutaneous metastatic nodules over the trunk. After perfusion of the limb, most of the metastatic lesions on the trunk disappeared. Another patient had extensive in-transit metastasis of the lower extremity. All lesions disappeared after hyperthermic perfusion. Seventeen months later, however, metastasis to a distant lymp node developed so the node was removed. The patient remains clinically free of disease 5 years after excision of the distant metastasis.

These clinical observations have been supported by experimental work in our laboratory. When cultured cells of human melanoma are exposed to autologous plasma or to both, the cytotoxic activity of the plasma and lymphocytes increases dramatically after hyperthermic perfusion. An example of this is shown in Figure 2. After hyperthermic perfusion which resulted in destruction of a large volume of melanoma in the extremity, the cytotoxicity of both plasma and lymphocytes rose within a few days to a high peak and then fell after a short time. The patient returned in 3 months with a small amount of residual tumor in the limb and was perfused again. The increase in cytotoxicity following the second perfusion was proportionately smaller.

The more we worked with heated perfusion, the more we were convinced that heat enhanced the antitumor effects of melphalan against melanoma. As this evidence began to mount, we attempted to transfer this regional concept of treatment to a systemic one. Seven years ago, we began using mixed bacterial toxins intravenously for the primary purpose of inducing fever. Once the patient's fever reached its maximum level,

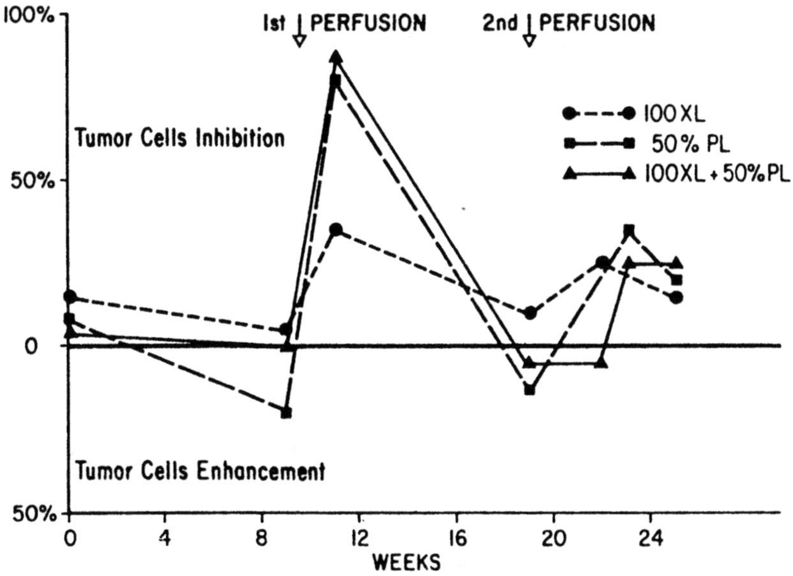

Fig. 2. Effects of hyperthermic perfusion upon cytotoxic activity of plasma and lymphocytes against cultured melanoma from the same patient. (From STEHLIN et al. (17))

standard amounts of dimethyl triazeno imidazole carboxamide were administered. The regimen was repeated on a daily basis for 10 days, and the course of therapy was administered once every 3 months. In the beginning, this therapeutic approach was used only for patients with stage IV disease. Recently, because of encouraging results, it has been extended to include patients with advanced but potentially curable melanoma. Not enough time has elapsed to evaluate this program properly, but it is our distinct impression that lives are being prolonged.

4. CONCLUSIONS

Our studies show that the malignant melanoma cell in human beings is more sensitive to the lethal effect of heat that its norman counterpart, the melanocyte. Malignant melanoma of the extremities presents unique problems; at times local control can be extremely difficult. The addition of heat to regional perfusion with melphalan has dramatically improved the objective response of melanoma. Complications rise as the temperature and duration of perfusion increase. These risks must be weighed carefully against the volume and extent of tumor. A total of 185 hyperthermic perfusions have been performed on 165 patients. When done with meticulous attention to details, this procedure is accompanied by minimal morbidity and mortality. Hyperthermic perfusion is currently the best treatment for recurrent melanoma of the extremities and has almost eliminated the necessity for amputation. Perfusion is recommended as a prophylactic measure for the more deeply invasive primary lesions. It reduces the incidence of regional recurrence. A retrospective statistical analysis of survival rates of patients treated with non-heated and heated perfusion for recurrent melanomas, stage IIIA, was conducted.

If the experience of the heated group continues, which from a clinical standpoint appears likely, then a striking advantage of heated perfusion over nonheated perfusion will be demonstrated. This superiority in survival rates for the heated group is now 3 to 1 or 300%. The most reasonable explanation for the improvement in survival time of patients in stage IIIA is stimulation of the immune response. As a result of our experience with heated perfusion of limbs, we are investigating the possibility that systemic hyperthermia may enhance the antitumor effects of various chemotherapeutic agents on melanoma.

5. SARCOMAS

Soft tissue sarcomas of the extremities present unique problems both from the standpoint of local-regional control and of distant metastasis. The difficulties in local control have prompted surgeons to use radical amputation as the treatment of choice for many of these patients. The loss of a limb, unfortunately, is followed by distant metastasis in a significant number of these patients (7). We have attempted to modify both the local and the distant behavior of this disease by a multidisciplinary approach utilizing hyperthermic perfusion and, when possible, irradiation and radical local excision. Salvage of the extremity has been a principal objective in every case; no patient has received amputation as a primary form of therapy. Therefore, contrary to other published reports on this subject, there is no case selection here insofar as the feasibility of amputation as a primary form of therapy is concerned.

During the past 8 years, we have treated 37 patients with soft tissue sarcomas of the extremities, all of whom were treated with a multidisciplinary approach utilizing hyperthermic perfusion and, when possible, external irradiation with radial local excision.

A. Technique of Hyperthermic Perfusion

The technique used is the same described before for the treatment of melanomas except for the choice of chemotherapeutic agents. Melphalan in combination with aetinomycin D was used for 33 of 37 patients; 1 was treated with melphalan alone and 3 were treated with nitrogen mustard alone (18).

B. Additional Treatment

As soon as the tissue reaction following perfusion has subsided, usually within 10 days, radiation therapy is begun. If radical local excision is anticipated, a preoperative course of 3000 rads T.D. in 3 weeks is given, using parallel opposing ports with the cobalt-60 teletherapy unit. Generous fields which may extend from joint to joint are employed. The postoperative course of irradiation is added if residual disease is found at the time of surgery. Then an additional 3000 rads T.D. in 3 weeks is given with reduced fields limited to the region of the surgical clips. If surgical excision is not anticipated, a total of about 6000 rads T.D. in 6 weeks is given.

When excision is feasible, it is delayed for 8-12 weeks following completion of radiation therapy. Meticulous attention is directed toward securing margins of normal tissue surrounding the tumor. This may involve the sacrifice of adjacent muscles. Though often technically simple, enucleation of the tumor through its pseudocapsule must be avoided because of the high incidence of local recurrence. For those lesions on the distal extremity, a proximal tourniquet provides a bloodless field for excision. After excision, surgical clips are applied at the points where the risk of local recurrence appears most likely. This facilitates localization of additional irradiation postoperatively to a smaller field.

C. Major Complications

Two of the 37 patients died in the immediate postoperative period; 1, a 74-year old woman, died from a myocardial infarction; the other, a 63-year old man, died from complications related to bleeding stress ulcers which followed an above-knee amputation for tissue necrosis. Perfusion resulted in tissue necrosis which required amputation in 1 additional patient. Both patients requiring amputation postoperatively had tumors for which amputation would have been mandatory under ordinary circumstances. A third patient, with periosteal fibrosarcoma of the femur, developed a nonhealing fracture attributed to radiation therapy. Following internal fixation, osteomyelitis necessitated amputation 28 months after perfusion. Most of these complications occured in the early phases of this work. As more experimence was gained, the complications were reduced to a minimum.

D. Classification of Sarcomas

From March 1967 to March 1975, 37 patients with soft tissue sarcomas of the extremities have been treated. The histopathologic types are listed in Table 3. Seven of the tumors were on the upper extremity and 30 on the lower. Sixteen of the patients were male and 21 female. The ages of the patients ranged from 7 to 74 years, with the average being 42.6 years.

Table 3. Classification of primary tumors

Liposarcoma	9	Leiomyosarcoma	2
Synovial sarcoma	7	Kaposi sarcoma	2
Rhabdomyosarcoma	6	Malignant schwannoma	1
Fibrosarcoma	4	Lymphangiosarcoma	1
Alveolar soft part sarcoma	3	Unclassified sarcoma	2
Total			37

From STEHLIN et al. (18).

E. Outline of Treatment

All 37 patients received hyperthermic perfusion. Seventeen of the 37 patients received perfusion, radiation therapy, and delayed radial local excision of either the primary tumor of the site of its origin. Nine of the 37 patients had undergone excisional biopsy of either the primary tumor or local recurrence prior to being sent to us for definitive therapy. Here they received perfusion plus postoperative irradiation. Seven of the 37 patients received perfusion followed by radiation therapy: 3 had tumors whose locations precluded excision short of amputation and 1 had diffuse lymphangiosarcoma of the upper extremity.

Once the diagnosis is confirmed by biopsy, hyperthermic perfusion is performed. The temperatures of the blood, skin, and muscles are monitored continously with thermistor probes. If tumor is grossly palpable, an additional probe is inserted into it. Heat is applied internally by raising the temperature of the perfused blood to 110°F (43.3°C). External heat is applied by wrapping the extermity with Aquamatic K-Pads. When the temperatures of the skin and muscles reach 102.0-104.0°F (38.8-40.0°C), the cytotoxic agents are injected into the venous line. Melphalan in combination with actinomycin D was used for 33 of 37 patients; 1 was treated with melphalan alone and 3 were treated with nitrogen mustard alone. Once the drug is introduced, the temperatures are monitored continously and adjustments are made as necessary to keep them at the desired levels. The temperature and duration of perfusion depend on the size of the sarcoma and the condition of the patient's vascular tree. The incidence of serious injury to normal tissues is directly proportional to the amount of drug used, the temperature reached in the tissues, the duration of the perfusion, and the condition of the peripheral vessels. We have set an upper limit of 2 1/2 h of tourniquet use to avoid damge to the nerves.

6. RESULTS

The effectiveness of treatment may be evaluated according to survival rates and incidence of local recurrence. The 5-year survival rate of all patients as projected by the Berkson-Gage method was 61.9%.

Following definitive treatment, 6 of the 37 patients had recurrent or metastatic sarcoma in the affected limb. In 4 of the 6, metastasis in the involved extremity appeared only as a late manifestation of widespread disseminated cancer and required no local therapy. The remaining 2 patients had extensive synovial sarcomas of the foot. Amputation was necessary for local recurrence 13 and 21 months after perfusion and irradiation. Of the 6 patients whose disease was not controlled locally, 5 died of cancer and the 6th is living with widespread sarcoma. Five additional patients died of cancer. Three of these were considered curable at the time of perfusion; 2 had distant metastatis when first treated here and were perfused for palliative purposes.

7. DISCUSSION

Two points in diagnosis deserve special emphasis. The first is the necessity on the part of the physician to recognize any painless lump as potentially malignant. The second concerns the late recognition of sarcoma of the thigh, as evidenced by the enormous size of some of them when first diagnosed. These tumors usually arise within the confines of the tight sleeve of fascia lata. Rather than presenting as a discrete lump, they often produce diffuse enlargement of the thigh, rendering diagnosis more difficult. Although some of these patients repeatedly sought medical attention because of vague discomfort in the thigh, the actual recognition by the physician of the presence of a mass was delayed for months.

A combined approach utilizing regional chemotherapy, irradiation, and surgical excision offers some very attractive features. Isolation perfusion permits massive concentration of cytotoxic agents within an extremity, something not possible with systemic chemotherapy. Although normothermic perfusion resulted in good responses for some sarcomas, a dramatic increase in the objective responses occurred in 1967 when heat was added. We believe that hyperthermic perfusion enhances the effectiveness of radiation therapy. Clinical fluctuation of some of these large masses has been noted as early as 3-4 days after perfusion. Following perfusion, irradiation, and delayed excision, some of these large tumors have shown microscopic and laboratory evidence of complete destruction. This complete destruction cannot be obtained by irradiation alone at a dosage of 3000 rads. To the contrary, many of these tumors have been traditionally considered radioresistant.

In evaluating the effectiveness of this combined therapy insofar as local control of the sarcoma is concerned, 3 patients must be omitted: the 2 who died postoperatively and the 1 who required amputation because of tissue necrosis resulting from perfusion. That this multidisciplinary therapy is effective is evidenced by the fact that only 2 of the remaining 34 patients (5.9%) developed local recurrence for which further surgery was necessary. This figure becomes even more meaningful when one realizes that there has been no case selection, and that 8 of these patients were referred to us because of local recurrence following surgical excision. There can be no doubt that this therapeutic approach has saved many useful extremities and has reduced to a minimum the necessity for amputation.

Any meaningful comparison of reported survival rates for soft tissue sarcomas is impossible without knowledge of specific parametes, including size of tumors and histopathologic types. To our knowledge, our 62% projected 5-year survival is as good as any reported series. Based on this evidence, the early fears that regional chemotherapy and delayed excision would lower survival rates are unfounded. On the contrary, destruction of sarcoma within the limb followed by delayed excision has one potential advantage which may improve surgical rates. The possibility exists that the breakdown products following destruction of tumor are absorbed into the systemic circulation and may stimulate the immune response of the patient against his own tumor. Much more work needs to be done before a definitive answer to this hypothesis is obtained.

8. CONCLUSIONS

1. Thirty-seven patients with soft tissue sarcomas of the extermities have been treated with hyperthermic perfusion, and, when possible, radiation therapy and radical local excision.

2. Despite the fact that these tumors arise on the extremity, diagnosis is often delayed.

3. With meticulous attention to all details, hyperthermic perfusion can be performed with a minimum of morbidity and mortality.

4. The combined use of hyperthermic perfusion, radiation therapy, and delayed excision has greatly reduced the necessity for amputation.

5. Many useful limbs, heretofore not considered salvageable, can be saved with this method of treatment.

6. The possibility that destruction of tumor within the extremity followed by delayed excision enhances the patient's immune response is being investigated.

REFERENCES

1. CAVALIERE, R., CIOCATTO, E.L., GIOVANELLA, B.D., HEIDELBERGER, C., JOHNSON, R.O., MARGOTTINI, M., MONDOVI, B., MORICCA, G., ROSSI-FANELLI, A.: Selective heat sensitivity of cancer cells. Cancer 20, 1351 (1967)
2. CLARK, W.H., FROM, L.: The histogenesis and biological behavior of primary human malignant melanomas of the skin. Cancer Res. 29, 29 (1969)
3. CREECH, I., Jr., KREMENTZ, E.T., RYAN, R.F., REEMTSMA, K., WINBLAD, J.N.: Experiences with isolation-perfusion technics in the treatment of cancer. Ann. Surg. (Chicago) 149, 627 (1959)
4. CREECH, I., Jr., RYAN, R.F., KREMENTZ, E.T.: Treatment of melanoma by isolation-perfusion technique. J. Amer. med. Ass. 169, 339 (1959)
5. CREECH, O., Jr., KREMENTZ, E.T., RYAN, R.F., REEMTSMA, K., WINBLAD, J.N., ELLIOT, J.L.: Treatment of cancer by perfusion. Arch. Surg. (Chicago) 79, 963 (1959)
6. CREECH, O., Jr., KREMENTZ, E.T., RYAN, R.F., WINBLAD, J.N.: Chemotherapy of cancer: Regional perfusion utilizing an extra-corporeal circuit. Ann. Surg. 148, 616 (1958)
7. FINE, G., OHORODNIK, J.M., HORN, R.C., Jr.: Soft-tissue sarcomas: Their clinical behavior and course and influencing factors. In: Seventh National Cancer Conference Proceedings (1972), Philadelphia: J.B. Lippincott Co., 1973
8. KLOPP, C.T.: Fractionated intra-arterial cancer chemotherapy with methyl bis amine hydrochloride: A preliminary report. Ann. Surg. 132, 811 (1950)
9. KLOPP, C.T., BATEMAN, J., BERY, N., ALFORD, C., WINSHIP, T.: Fractionated regional cancer chemotherapy. Cancer Res. 10, 229 (1950)
10. McNEER, G., CANTIN, J.: Local failure in the treatment of melanoma. Amer. J. Roentgenol. 99, 791 (1967)
11. STEHLIN, J.S., Jr.: Hyperthermic perfusion with chemotherapy for cancer of the extremities. Surg. Gynec. Obstet. 129, 305 (1969)
12. STEHLIN, J.S., Jr., CLARK, R.L.: Chemotherapy for cancer by regional perfusion. Arch. Surg. (Chicago) 83, 196 (1961)
13. STEHLIN, J.S., Jr., CLARK, R.L., Jr., DEWEY, W.C.: Continuous monitoring of leakage during perfusion. Arch. Surg. (Chicago) 83, 943 (1961)
14. STEHLIN, J.S., Jr., CLARK, R.L., Jr., WHITE, E.C., HEALEY, J.E., Jr., DEWEY, W.C., BEERSTECHER, S.: The leakage factor in regional perfusion of chemotherapeutic agents. Arch. Surg. (Chicago) 80, 934 (1960)
15. STEHLIN, J.S., Jr., CLARK, R.L., Jr., WHITE, E.C., SMITH, J.L., Jr.: Malignant melanoma of the extremities: Experiences with conventional therapy; A new surgical and chemotherapeutic approach with regional perfusion. Cancer (Philad.) 13, 55 (1960)

16. STEHLIN, J.S., Jr., CLARK, R.L., Jr., WHITE, E.C., SMITH, J.L., Jr., GRIFFIN, A.C., JESSE, R.H., Jr., HEALEY, J.E., Jr.: Regional chemotherapy for cancer: Experiences with 116 perfusions. Ann. Surg. 151, 605 (1960)
17. STEHLIN, J.S., Jr., GIOVANELLA, B.C., de IPOLYI, P.D., MUENZ, L.R., ANDERSON, R.F.: Results of hyperthermic perfusion for melanoma of the extremities. Surg. Gynec. Obstet. 140, 338 (1975)
18. STEHLIN, J.S., Jr., de IPOLYI, P.D., DIOVANELLA, B.C., GUTIERREZ, A.E., ANDERSON, R.F.: Soft tissue sarcomas of the extremity, multidisciplinary therapy employing hyperthermic perfusion. Amer. J. Surg. 130, 643 (1975)
19. STEHLIN, J.S., Jr., SMITH, J.L., JING, B., SHERRIN, D.: Melanomas of the extremities complicated by in-transit metastases. Surg. Gynec. Obstet. 122, 3 (1966)

STEHLIN, J.S., Jr., CLARK, R.L., Jr., WHITE, E.C., SMITH, J.L., Jr., GRIFFIN, A.C., JESSE, R.H., Jr., HEALEY, J.E., Jr.: Regional chemotherapy for cancer: Experiences with 116 perfusions. Ann. Surg. 151, 605 (1960)
17. STEHLIN, J.S., Jr., GIOVANELLA, B.C., de IPOLYI, P.D., MUENZ, L.R., ANDERSON, R.F.: Results of hyperthermic perfusion for melanoma of the extremities. Surg. Gynec. Obstet. 140, 338 (1975)
18. STEHLIN, J.S., Jr., de IPOLYI, P.D., GIOVANELLA, B.C., GUTIERREZ, A.E., ANDERSON, R.F.: Soft tissue sarcomas of the extremity. Multidisciplinary therapy employing hyperthermic perfusion. Amer. J. Surg. 130, 643 (1975)
19. STEHLIN, J.S., Jr., SMITH, J.L., JING, B., SHERRIN, D.: Melanomas of the extremities complicated by in-transit metastases. Surg. Gynec. Obstet. 122, 3 (1966)

Subject Index

Recent Results in Cancer Research

Sponsored by the Swiss League against Cancer. Editor in Chief: P. Rentchnick, Genève

For information about Vols. 1—9, please contact your bookseller or Springer-Verlag.

10 NELSON, R. S.: Radioactive Phosphorus in the Diagnosis of Gastrointestinal Cancer.
11 FREEMAN, R. G. and J. M. KNOX: Treatment of Skin Cancer.
12 LYNCH, H. T.: Hereditary Factors in Carcinoma.
13 Tumours in Children, 2nd Edition. Edited by H. B. MARSDEN and J. K. STEWARD.
14 ODARTCHENKO, N.: Production Cellulaire Erythropoïétique.
15 SOKOLOFF, B.: Carcinoid and Serotonin.
16 JACOBS, M. L.: Malignant Lymphomas and Their Management.
17 Normal and Malignant Cell Growth. Edited by R. J. M. FRY, M. L. GRIEM, and W. H. KIRSTEN (Symposium).
18 ANGLESIO, E.: The Treatment of Hodgkin's Disease.
19 BANNASCH, P.: The Cytoplasm of Hepatocytes during Carcinogenesis. Electron- and Lightmicroscopical Investigations of the Nitrosomorpholine-intoxicated Rat Liver.
20 Rubidomycin. A new Agent against Cancer. Edited by J. BERNARD, R. PAUL, M. BOIRON, C. JACQUILLAT, and R. MARAL.
21 Scientific Basis of Cancer Chemotherapy. Edited by G. MATHÉ (Symposium).
22 KOLDOVSKÝ, P.: Tumor Specific Transplantation Antigen.
23 FUCHS, W. A., J. W. DAVIDSON, and H. W. FISCHER: Lymphography in Cancer. With contributions by G. JANTET and H. RÖSLER.
24 HAYWARD, J.: Hormones und Human Breast Cancer. An Account of 15 Years Study.
25 ROY-BURMAN, P.: Analogues of Nucleic Acid Components. Mechanisms of Action.
26 Tumors of the Liver. Edited by G. T. PACK and A. H. ISLAMI.
27 SZYMENDERA, J.: Bone Mineral Metabolism in Cancer.
28 MEEK, E. S.: Antitumour and Antiviral Substances of Natural Origin.
29 Aseptic Environments and Cancer Treatment. Edited by G. MATHÉ (Symposium).
30 Advances in the Treatment of Acute (Blastic) Leukemias. Edited by G. MATHÉ (Symposium).
31 DENOIX, P.: Treatment of Malignant Breast Tumors. Indications and Results.
32 NELSON, R. S.: Endoscopy in Gastric Cancer.
33 Experimental and Clinical Effects of L-Asparaginase. Edited by E. GRUNDMANN and H. F. OETTGEN (Symposium).
34 Chemistry and Biological Actions of 4-Nitroquinolin 1-Oxide. Edited by H. ENDO, T. ONO, and T. SUGIMURA.
35 PENN, I.: Malignant Tumors in Organ Transplant Recipients.
36 Current Concepts in the Management of Lymphoma and Leukemia. Edited by J. E. ULTMANN, M. L. GRIEM, W. H. KIRSTEN, and R. W. WISSLER (Symposium).
37 CHIAPPA, S., R. MUSUMECI, and C. USLENGHI: Endolymphatic Radiotherapy in Malignant Lymphomas. With contributions by G. BONADONNA, B. DAMASCELLI, G. FAVA, F. PIZZETTI, U. VERONESI.
38 KOLLER, P. C.: The Role of Chromosomes in Cancer Biology.
39 Current Problems in the Epidemiology of Cancer and Lymphomas. Edited by E. GRUNDMANN and H. TULINIUS (Symposium).
40 LANGLEY, F. A. and A. C. CROMPTON: Epithelial Abnormalities of the Cervix Uteri.
41 Tumours in a Tropical Country. A Survey of Uganda (1964—1968). Edited by A. C. TEMPLETON.
42 Breast Cancer: A Challenging Problem. Edited by M. L. GRIEM, E. V. JENSEN, J. E. ULTMANN, and R. W. WISSLER (Symposium).
43 Nomenclature, Methodology and Results of Clinical Trials in Acute Leukemias. Edited by G. MATHÉ, P. POUILLART, L. SCHWARZENBERG (Symposium).

44 Special Topics in Carcinogenesis. Edited by E. GRUNDMANN (Symposium).
45 KOLDOVSKÝ, P.: Carcinoembryonic Antigens.
46 Diagnosis and Therapy of Malignant Lymphoma. Edited by K. MUSSHOFF (Symposium).
47 Investigation and Stimulation of Immunity in Cancer Patients. Edited by G. MATHÉ and R. WEINER (Symposium).
48 Platinum Coordination Complexes in Cancer Chemotherapy. Edited by T. A. CONNORS and J. J. ROBERTS (Symposium).
49 Complications of Cancer Chemotherapy. Edited by G. MATHÉ and R. K. OLDHAM (Symposium).
50 Cancer Registry. Edited by E. GRUNDMANN and E. PEDERSEN (Symposium).
51 Gliomas. Current Concepts in Biology, Diagnosis and Therapy. Edited by J. HEKMATPANAH (Symposium).
52 The Ambivalence of Cytostatic Therapy. Edited by E. GRUNDMANN and R. GROSS (Symposium).
53 A. CLARYSSE, Y. KENIS, and G. MATHÉ: Cancer Chemotherapy.
54 Malignant Bone Tumors. Edited by E. GRUNDMANN.
55 MATHÉ, G.: Cancer Active Immunotherapy, Immunoprophylaxis, and Immunorestoration.
56 Lymphocytes, Macrophages, and Cancer. Edited by G. MATHÉ, I. FLORENTIN, and M.-C. SIMMLER (Symposium).
57 Breast Cancer: A Multidisciplinary Approach. Edited by G. ST. ARNEAULT, P. BAND, and L. ISRAËL (Symposium).
58 B. S. SCHOENBERG: Multiple Primary Malignant Neoplasms.
59 Selective Heat Sensitivity of Cancer Cells. Edited by A. ROSSI-FANELLI, R. CAVALIERE, B. MONDOVI, and G. MORICCA.
60 Tumors of the Male Genital System. Edited by E. GRUNDMANN and W. VAHLENSIECK (Symposium).

Special Supplement: Biology of Amphibian Tumors. Edited by M. MIZELL.

MIX
Papier aus verantwortungsvollen Quellen
Paper from responsible sources
FSC® C105338

If you have any concerns about our products,
you can contact us on
ProductSafety@springernature.com

In case Publisher is established outside the EU,
the EU authorized representative is:
Springer Nature Customer Service Center GmbH
Europaplatz 3, 69115 Heidelberg, Germany

Printed by Libri Plureos GmbH
in Hamburg, Germany